Springer Theses

Recognizing Outstanding Ph.D. Research

Aims and Scope

The series "Springer Theses" brings together a selection of the very best Ph.D. theses from around the world and across the physical sciences. Nominated and endorsed by two recognized specialists, each published volume has been selected for its scientific excellence and the high impact of its contents for the pertinent field of research. For greater accessibility to non-specialists, the published versions include an extended introduction, as well as a foreword by the student's supervisor explaining the special relevance of the work for the field. As a whole, the series will provide a valuable resource both for newcomers to the research fields described, and for other scientists seeking detailed background information on special questions. Finally, it provides an accredited documentation of the valuable contributions made by today's younger generation of scientists.

Theses are accepted into the series by invited nomination only and must fulfill all of the following criteria

- They must be written in good English.
- The topic should fall within the confines of Chemistry, Physics, Earth Sciences, Engineering and related interdisciplinary fields such as Materials, Nanoscience, Chemical Engineering, Complex Systems and Biophysics.
- The work reported in the thesis must represent a significant scientific advance.
- If the thesis includes previously published material, permission to reproduce this must be gained from the respective copyright holder.
- They must have been examined and passed during the 12 months prior to nomination.
- Each thesis should include a foreword by the supervisor outlining the significance of its content.
- The theses should have a clearly defined structure including an introduction accessible to scientists not expert in that particular field.

More information about this series at http://www.springer.com/series/8790

Lavinia Heisenberg

Theoretical and Observational Consistency of Massive Gravity

Doctoral Thesis accepted by
the University of Geneva, Switzerland

Springer

Author
Dr. Lavinia Heisenberg
Royal Institute of Technology
Stockholm
Sweden

Supervisor
Prof. Claudia de Rham
Case Western Reserve University
Cleveland, OH
USA

ISSN 2190-5053 ISSN 2190-5061 (electronic)
Springer Theses
ISBN 978-3-319-18934-5 ISBN 978-3-319-18935-2 (eBook)
DOI 10.1007/978-3-319-18935-2

Library of Congress Control Number: 2015940978

Springer Cham Heidelberg New York Dordrecht London

Printed on acid-free paper

Springer International Publishing AG Switzerland is part of Springer Science+Business Media (www.springer.com)

Parts of this thesis have been published in the the following journal articles:

This doctoral thesis covers a detailed presentation of the scientific results published in the following articles. The results are not presented in chronological order of their publication but rather in the logically most comprehensible way. The thesis also contains unpublished results, which have been pointed out at the appropriate place.

1. C. de Rham, L. Heisenberg, R.H. Ribeiro:
   Quantum Corrections in Massive Gravity:
   Phys. Rev. D 88, 084058 (2013)
2. C. de Rham, G. Gabadadze, L. Heisenberg, D. Pirtskhalava:
   Non-Renormalization and Naturalness in a Class of Scalar-Tensor Theories:
   Phys. Rev. D 87, 085017 (2013)
3. P. de Fromont, C. de Rham, L. Heisenberg, A. Matas:
   Superluminality in the Bi- and Multi- Galileon:
   JHEP 1307 (2013) 067
4. C. de Rham, L. Heisenberg:
   Cosmology of the Galileon from Massive Gravity:
   Phys. Rev. D 84, 043503 (2011)
5. C. Burrage, C. de Rham, L. Heisenberg:
   De Sitter Galileon:
   JCAP 05 (2011) 025
6. C. de Rham, G. Gabadadze, L. Heisenberg, D. Pirtskhalava:
   Cosmic Acceleration and the Helicity-0 Graviton:
   Phys. Rev. D 83, 103516 (2011)

Additional publications related to this thesis.

1. C. Burrage, C. de Rham, L. Heisenberg, A.J. Tolley:
   Chronology Protection in Galileon Models and Massive Gravity:
   JCAP 07 (2012) 004
2. J.B. Jimenez, R. Durrer, L. Heisenberg, M. Thorsrud:
   Stability of Horndeski Vector-Tensor Interactions:
   JCAP 1310 (2013) 064

The Road Not Taken

TWO roads diverged in a yellow wood,
And sorry I could not travel both
And be one traveler, long I stood
And looked down one as far as I could
To where it bent in the undergrowth;

Then took the other, as just as fair,
And having perhaps the better claim,
Because it was grassy and wanted wear;
Though as for that the passing there
Had worn them really about the same,

And both that morning equally lay
In leaves no step had trodden black.
Oh, I kept the first for another day!
Yet knowing how way leads on to way,
I doubted if I should ever come back.

I shall be telling this with a sigh
Somewhere ages and ages hence:
Two roads diverged in a wood, and I –
I took the one less traveled by,
And that has made all the difference.

—Robert Frost

Dedicated to the endless memories of Rojdén and Zera.

Supervisor's Foreword

It is with great pride that I look back on the 5 years since Lavinia Heisenberg started her Ph.D. at Geneva University. Committed to work on theoretical aspects of Cosmology, Lavinia joined the field of modified gravity as a tool to explain the late-time acceleration of the Universe and tackle the Cosmological Constant problem.

The Cosmological Constant problem, or why the vacuum energy of particle physics is so small, is a longstanding puzzle existing now for more than seven decades. The recent realizations that our Universe is accelerating and that this acceleration is very well fitted by a Cosmological Constant has revived the old Cosmological Constant problem as well as the search for alternative theories.

At the time when Lavinia started her thesis research, most consistent models of modified gravity in the infrared were driven by higher-dimensional setups and the "Galileon" scalar field had just emerged as an effective description of such theories. During the first year of Lavinia's Ph.D., the field of modified gravity underwent a remarkable transition and Lavinia bonded with it in an exceptional way.

First, massive gravity (a theory of gravity where the particle responsible for the gravitational force has a mass) was derived as its own consistent theory, an effort which was previously thought to be impossible. From there, theories of bi-gravity and multi-gravity emerged as new ways to understand gravity. These new theories required the development of many new techniques that were so far lacking in the literature and Lavinia literally took the role of a forerunner in developing many of them.

Second, many properties of the Galileon scalar fields were still to be explored, and throughout her research, Lavinia has been devoted to understanding their theoretical and phenomenological properties in-depth. From extending the Galileon symmetry to other maximally symmetric spacetimes, to understanding their quantum behavior and the implications of their superluminal propagation, Lavinia's work has pushed our understanding of these fields to a new level. These studies became all the more relevant after the realization that the helicity-0 mode of the graviton in massive gravity and bi- or multi-gravity also behaves as a Galileon scalar field in a certain limit.

Motivated by this observation, Lavinia was able to establish the existence of 'degravitating' solutions in massive gravity where a large cosmological constant could effectively be 'eaten' by the graviton mass. Although these solutions cannot degravitate an arbitrarily large cosmological constant without also exciting the helicity-0 mode at an undesirably low scale, this framework still provides a proof-of-principle for circumventing Weinberg's no-go theorem when it comes to tackling the Cosmological Constant problem.

Related to the late-time acceleration of the Universe, Lavinia's thesis also shows how the helicity-0 mode of the graviton can itself 'self-accelerate' the expansion of the Universe without the need for any dark energy.

Based on these results, Lavinia was then able to describe how to derive a proxy theory for massive gravity which captures the essential ingredients of massive gravity without many of its complications. This proxy theory was shown to be a special case of Horndeski Scalar-Tensor theory, which has independently enjoyed a great deal of attention recently. Within the context of this proxy theory, Lavinia was then able to capture the essence of the self-accelerating solution in more depth, taking into account the cosmological history of the Universe and presenting the first elements toward a study of structure formation. A subsequent analysis has established that the self-accelerating solution can only be a transient regime within that proxy theory and can never be a late time attractor.

Furthermore, motivated by the derivation of the proxy theory and their relation with Horndeski scalar–tensor theories, Lavinia also generalized the Horndeski framework to vector fields and established the first most general consistent vector–tensor theory, as well as exploring the phenomenological implications for Cosmology.

Whether it is in massive gravity, in Galileon theories or in more general Horndeski theories, the presence of superluminal modes has been a source of concern and arguments for more than a decade. When Lavinia started her thesis research, it had been established that a certain class of bi-Galileon theories could avoid any superluminality issues while still exhibiting a consistent self-accelerating solution, as well as an active Vainshtein mechanism. This realization attracted much interest and theories of multi-gravity or of gravity embedded in more than one extra dimension were investigated with the hope of avoiding the superluminality issues, since they behave as multi-Galileon theories in some limit. Thanks to Lavinia's work, this issue was re-explored and it was found that the presence of superluminalities is actually unavoidable in all of these classes of theories, even though closed-time-like curves are impossible within the regime of validity of such theories.

When the new theories of modified gravity came out (massive gravity, bi- and multi-gravity, as well as their extensions), two of the most pressing questions were related to their quantum stability and their cosmological phenomenology. It is fair to say that Lavinia's work represented major advances in both directions. The quantum stability was first established in the Galileon limit at all loops, and then beyond that limit at the one-loop level. While a full analysis is yet to be performed, the arguments presented in this thesis go a long way toward understanding the technical naturalness of the graviton mass and the stability of the allowed interactions in these types of theories.

On the cosmological side, Lavinia's thesis has established many different techniques to explore the phenomenology of these theories, both at early and late-time, and combining these results with data analysis from different cosmological probes. In particular, Lavinia was able to generalize the coupling to matter in these types of theories and provide new directions to study their cosmology.

Much of Lavinia's thesis has now taken on a life of its own and the issues are being explored further. There is no doubt, however, that it gathers together many different facets of the recent developments in the field of massive gravity and will serve as a valuable reference for the field.

Cleveland, USA
December 2014

Prof. Claudia de Rham

Abstract

This doctoral thesis encompasses a detailed study of phenomenological as well as theoretical consequences derived from the existence of a graviton mass within the ghost-free theory of massive gravity, the de Rham-Gabadadze-Tolley (dRGT) theory, which incorporates a two-parameter family of potentials. In this thesis we pursue to test the physical viability of the theory. To start with, we have put constraints on the parameters of the theory in the decoupling limit based on purely theoretical grounds, like classical stability in the cosmological evolution. Hereby, we were able to construct self-accelerating solutions which yield similar cosmological evolution to a cosmological constant. Furthermore, we studied the degravitating solutions, which enables us to screen an arbitrarily large cosmological constant in the decoupling limit. Nevertheless, conflicts with observations push the allowed value of the vacuum energy to a very low value rendering the found degravitating solution phenomenologically not viable for tackling the old cosmological constant problem. Next, we constructed a proxy theory to massive gravity from the decoupling limit resulting in non-minimally coupled scalar–tensor interactions as an example of a subclass of Horndeski theories. We explored the self-accelerating and degravitating solutions in this proxy theory in analogy to the decoupling limit and extended the analysis by studying the change in the linear structure formation.

Furthermore, Galileon models are a class of effective field theories that naturally arise in the decoupling limit of theories of massive gravity. We show that the existence of superluminal propagating solutions for multi-Galileon theories is an unavoidable feature.

Finally, we addressed the natural question of whether the introduced parameters in the theory are subject to strong renormalization by quantum loops. Starting with the decoupling limit we have shown how the non-renormalization theorem protects the graviton mass from quantum corrections. Beyond the decoupling limit the quantum corrections are proportional to the graviton mass, proving its technical naturalness in an explicit realization of 't Hooft's naturalness argument. Moreover, we pushed the analysis beyond the decoupling limit by studying the stability of the graviton potential when including matter and graviton loops. One-loop matter corrections contribute a cosmological constant term leaving the potential unaffected. On the contrary, the one-loop contributions from the gravitons destabilize the special structure of the potential. Nevertheless, we showed that even in the case of large background configuration, the Vainshtein mechanism redresses the one-loop effective action so that the detuning remains irrelevant below the Planck scale.

Acknowledgments

I would like to take this chance to raise my deepest gratitude to my thesis advisor, Claudia de Rham, for her impeccable supervision, endless support, patience, and stimulating guidance. I am very thankful that she gave me the opportunity to work with her in an active and interesting research field. I feel extremely fortunate to have had such an active and engaged advisor. I am even more thankful that she stayed committed to my formation as a researcher when she was offered a permanent position at Case Western Reserve University (CWRU) in Cleveland. I especially appreciate the effort that she has made all the time to get things done properly and that my training was not neglected. We have managed to work together on several projects over distance and she made it possible for me to visit her at CWRU for an extended period of time. It is also an appropriate place to express my gratitude to all my colleagues and friends at CWRU who made my stay abroad enjoyable, specially Emanuela Dimastrogiovanni, Matteo Fasiello, David Jacobs, Andrew Matas, Lucas Keltner, Raquel Ribeiro, and Amanda Yoho.

The major part of this thesis has been developed in the Department of Theoretical Physics at the University of Geneva, and thanks to all the facilities the university provided me, and under the Swiss national funding, I was able to attend a large number of conferences and schools. I was able to profit a lot with my visits to the USA and Japan and I am very thankful to the Swiss national funding for providing financial support. Furthermore, I am very thankful to two very special persons at the University of Geneva, to Ruth Durrer, who has been always very supportive and patient and to Michele Maggiore for very useful discussions. I would also like to thank my colleagues and friends at the University of Geneva, especially Jose Beltran Jimenez for the numerous interesting conversations and for being supportive, Guillermo Ballesteros for his persistent questionings which gave rise to so many interesting knowledgeable conversations, and Dani Figureroa for our sushi evenings.

I also would like to thank Bjoern Malte Schaefer and Matthias Bartelmann at the University of Heidelberg for staying interested to work with me on subjects unrelated to my Ph.D. research and for their hospitality each time I visited them in Heidelberg. Each visit resulted in so many fruitful discussions and valuable knowledge. Special thanks to my colleagues and friends at the institute for theoretical astrophysics in Heidelberg, especially to Jean-Claude Waizmann, Gero Juergens, and Christian Angrick.

Finally, I would like to thank Sara, Justus, and Eylem for their unconditional love and support and for always being there for me.

Contents

Part III Quantum Corrections in Massive Gravity

Part IV Summary

Chapter 1
Introduction

1.1 Field Theories in Cosmology

Physics studies Nature, its matter content and its evolution modeled by the laws of physics. Physicists like Newton, Galileo or Kepler successfully postulated the physical laws of the "every day physics scales", the theory of classical mechanics, which describes the motion of objects with velocities much smaller than the speed of light and with sizes much larger than atoms or molecules. Nevertheless, on the two edges of these scales, i.e. on very large or very small scales and at speeds close to the speed of light, the theory of classical mechanics breaks down. After a significant theoretical and observational efforts pioneered by physicists like Einstein, Planck, Heisenberg, Dirac or Schroedinger, the laws of physics were also extended towards these extreme scales, incorporating concepts of relativity and quantum mechanics to describe physics on atomic and galactic scales. As an outcome, physicists now describe the macroscopic and microscopic world by two simple standard models: the Standard Model of particle physics and the Standard Model of Big Bang cosmology. A dream of every physicist is the unification of these two standard models into a single ultimate "theory of everything" in a consistent way. Thanks to the advances in our understanding of many physical phenomena at a fundamental level, we are witnessing remarkable attempts towards this direction.

The Standard Model of particle physics unifies the electro-weak and strong interactions with an exquisite experimental success. It is a theory consisting of elementary and composite particles described by the robust framework of quantum field theory. In this picture, particles correspond to the excited states of an underlying physical field which can be created/annihilated by local operators given by the irreducible unitary representations of the Poincaré group (Weinberg 2005). It is the $SU(3) \times SU(2) \times U(1)$ gauge symmetry that defines the Standard model of particle physics. The group $SU(3)$ corresponds to the color gauge symmetry of Quantum Chromodynamics, whereas $SU(2) \times U(1)$ is the gauge symmetry of the electro-weak interaction and breaks down spontaneously to $U(1)$ through the Higgs mechanism, process thanks to which the elementary particles acquire their masses. This symme-

L. Heisenberg, *Theoretical and Observational Consistency of Massive Gravity*, Springer Theses, DOI 10.1007/978-3-319-18935-2_1

try breaks spontaneously down to SU(2) $\times$ U(1) which is the remaining symmetry of the electro-weak interactions and to SU(3) which is the symmetry of the quantum chromodynamics.

The Standard Model consists of fermions and bosons, which differ fundamentally from each other by their spin statistics. The twelve elementary fermions divide into six leptons (electron, muon, tau and the corresponding neutrinos) and six quarks (up, down, charm, strange, top and bottom). The twelve bosons (photon, $W^{\pm}$, Z, eight gluons) carry the strong, weak and electromagnetic forces. Interactions between the electrically charged particles mediated by the photons are successfully described by quantum electrodynamics, whereas the interactions between quarks (color charged) and gluons is described by quantum chromodynamics. The weak force mediated by the $W^{\pm}$, Z bosons is mathematically merged with the electromagnetic force via electroweak interaction. Last but not least the Higgs boson detected recently at LHC is the crucial ingredient to explain the masses of the particles. All these elementary particles come in with different masses and from a theoretical point of view it is essential to study the interactions of massless and massive spin-n particles. The natural starting point is to find the consistent Lagrangian of the classical field theories with the corresponding Hamiltonian being bounded from below. Once this non-trivial prerequisite is successfully fulfilled, then the classical fields can be quantized using different quantization techniques like canonical quantization or path integral quantization. Even though Standard Model of particle physics provides theoretical robustness of quantum fields and breathtaking experimental predictions, it is still far from being complete since it does not incorporate the theory of general relativity.

Cosmology has progressively developed from a philosophical to an empirical scientific discipline. Given the high precision achieved by the cosmological observations, cosmology is now a suitable arena to test fundamental physics. The challenging task of cosmology is to unite the physics of the large scale structures in the Universe with the physics of the small scale structures in order to describe the dynamics of the Universe successfully. Therefore, cosmology is highly multi-disciplinary and merges together concepts from general relativity, quantum mechanics, field theory, fluid mechanics and statistics. Furthermore, its interplay with high energy and particle physics facilitates the creation of synergies between these different fields.

Observations of the Cosmic Microwave Background (CMB), supernovae Ia (SNIa), lensing and Baryon Acoustic Oscillations (BAO) have led to the cosmological standard model which requires an accelerated expansion of the late Universe, driven by dark energy. The physical origin of the accelerated expansion is still a mystery. In the Standard Model of particles the detection of the missing fundamental particle, the Higgs boson, was a revolutionary event and an unexaggerated merit of Nobel-prize. In a similar way, the missing particles in the Standard Model of cosmology, like the graviton or dark matter, and the resolution to the puzzle of accelerated expansion and its origin would be as revolutionary. There are promising explanatory attempts which fall into three primary categories.

The first solution consists of considering a small cosmological constant λ with a constant energy density giving rise to an effective repulsive force between cosmological objects at large distances (Peebles and Ratra 2003). The Einstein-Hilbert action is invariant under general diffeomorphisms and a cosmological constant λ can be included to this action without breaking this symmetry.[1] If we assume that the cosmological constant corresponds to the vacuum energy density, then the theoretical expectations for the vacuum energy density caused by fluctuating quantum fields differ from the observational bounds on λ by up to 120 orders of magnitude. This giagantic mismatch between the theoretically computed high energy density of the vacuum and the low observed value has remained for decades as one of the most challenging puzzles in theoretical physics and is called the cosmological constant problem. Indeed the cosmological constant problem is a puzzle concerning both particle physics and cosmology, since it involves quantum field theory techniques applied to cosmology. One of the lines taken in this thesis will be trusting the result from particle physics and tackle the cosmological constant problem from the gravity side, although many of the techniques employed lie at the interface between particle physics and cosmology.

The second solution could for instance consist in introducing new dynamical degrees of freedom by invoking new fluids with negative pressure. Quintessence is one of the important representatives of this class of modifications. The acceleration is due to a scalar field whose kinetic energy is small in comparison to its potential energy, causing a dynamical equation of state with initially negative values (Doran et al. 2001). This class of theories might exhibit fine-tuning problems analogous to the cosmological constant.

Alternatively, the third solution would correspond to explaining the acceleration of the Universe by changing the geometrical part of Einstein's equations. In particular, weakening gravity on cosmological scales might not only be responsible for a late-time speed-up of the Hubble expansion, but could also tackle the cosmological constant problem. Such scenarios arise in infrared modifications of general relativity like massive gravity or in higher-dimensional frameworks, which will be summarized shortly in the following.

1.1.1 Infrared Modifications of GR

In this thesis we will consider the first and third categories, namely cosmological constant λ and modified gravity. We will particularly study the infra-red modifications of gravity. One of the important large scale modified theories of gravity in the higher dimensional picture is the Dvali-Gabadadze-Porrati (DGP) model (Dvali et al. 2000). In this braneworld model our Universe is confined to a three-brane embedded in a five-dimensional bulk (Fig. 1.1). On small scales, four-dimensional gravity is recovered due to an intrinsic Einstein Hilbert term given by the brane curvature, whereas on larger, cosmological scales gravity is systematically weaker as the graviton leaks into the extra dimension. The action of the DGP model is given by

[1] In fact, it must be included from an effective field theory point of view.

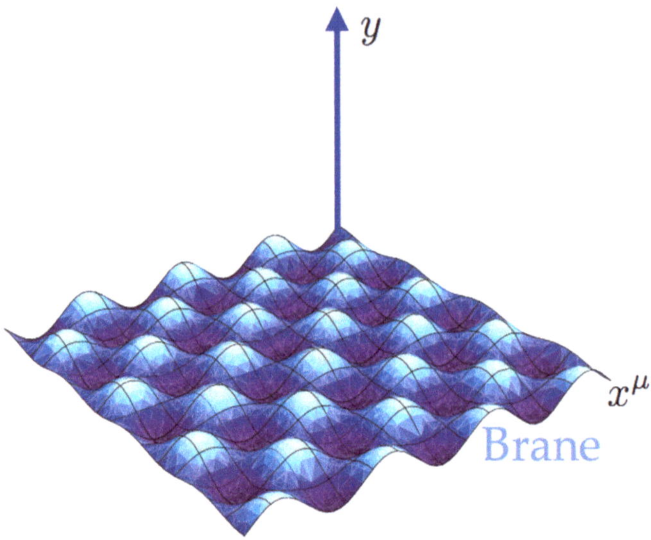

Fig. 1.1 Could our universe be just a part of higher dimensional space-time?

$$S_{DGP} = \frac{M_5^3}{2}\int d^5x\sqrt{-g_5}R_5 + \frac{M_{Pl}^2}{2}\int d^4x\sqrt{-g_4}R_4 + M_{Pl}^2\int d^4x\sqrt{-g_4}K \tag{1.1}$$

where M_{Pl} and M_5 respectively correspond to the fundamental Planck scales in the bulk and on the brane and K is the trace of the extrinsic curvature on the brane. Similarly, R_5 and R_4 are the corresponding Ricci scalars on the bulk and the brane respectively. The brane is positioned at $y = 0$ where y denotes the extra fifth dimension and x_μ are the four-dimensional coordinates. The crossover scale between 4- and 5-dimensional gravity is given by the ratio of these two Planck scales: $r_c = 1/m_c$ where $m_c = M_5^3/M_{Pl}^2$.

Using the principle of least action one obtains the modified Einstein equations

$$M_5^3 G_{ab}^{(5)} + M_{Pl}^2 G_{\mu\nu}^{(4)} \delta_a^\mu \delta_b^\nu \delta(y) = T_{\mu\nu}\delta_a^\mu \delta_b^\nu \delta(y). \tag{1.2}$$

where here $a, b = 0, \cdots, 4$ and $\mu, \nu = 0, \cdots, 3$. Being a fundamentally higher dimensional theory, the effective four-dimensional graviton on the brane carries five degrees of freedom, namely the usual helicity-2 modes, two helicity-1 modes and one helicity-0 mode. Whilst the helicity-1 modes typically decouple, the helicity-0 one can mediate an extra fifth force. In the limit $m_c \to 0$, one recovers General Relativity (GR) through the Vainshtein mechanism: The basic idea is to decouple the additional modes from the gravitational dynamics via nonlinear interactions of the helicity-0 mode of the graviton, Vainshtein (1972). As a result, at the vicinity of matter, the non-linear interactions for the helicity-0 mode become large and hence suppress its coupling to matter. This decoupling of the nonlinear helicity-0 mode is manifest in the limit where $M_{Pl} \to \infty$ and $m_c \to 0$ while the strong coupling scale $\Lambda_3 = (M_{Pl} m_c^2)^{1/3}$ is kept fixed. This limit enables a linear treatment of the usual helicity-2 mode of gravity while the helicity-0 mode π is described non-linearly, which is the so-called decoupling limit (Luty et al. 2003).

One of the successes of the DGP model is the existence of a self-accelerating solution, where the acceleration of the Universe is sourced by the graviton own degrees of freedom (more precisely its helicity-0 mode). Unfortunately that branch of solutions seems to be plagued by ghost-like instabilities (Deffayet et al. 2002; Koyama 2005; Charmousis et al. 2006), in the DGP model, but this issue could be avoided in more sophisticated setups, for instance including Gauss-Bonnet terms in the bulk (de Rham and Tolley 2006).

More recently, it has been shown that the decoupling limit of DGP could be extended to more general Galilean invariant interactions (Nicolis et al. 2009). This Galileon model relies strongly on the symmetry of the helicity-0 mode π: Invariance under internal Galilean and shift transformations, which in induced gravity braneworld models can be regarded as residuals of the 5-dimensional Poincaré invariance. These symmetries and the postulate of ghost-absence restrict the construction of the effective π Lagrangian. There exist only five derivative interactions which fulfill these conditions (1.83). From the five dimensional point of view these Galilean invariant interactions are consequences of Lovelock invariants in the bulk of generalized braneworld models, de Rham and Tolley (2010). Since their inception there has been a flurry of investigations related to self-accelerating de Sitter solutions without ghosts (Nicolis et al. 2009; Silva and Koyama 2009), Galileon cosmology and its observations (Chow and Khoury 2009; Khoury and Wyman 2009), inflation (Creminelli et al. 2010; Burrage et al. 2011; Mizuno and Koyama 2010; Hinterbichler and Khoury 2012), lensing (Wyman 2011), superluminalities arising in spherically symmetric solutions around compact sources (Hinterbichler et al. 2009), K-mouflage (Babichev et al. 2009), Kinetic Gravity Braiding (Deffayet et al. 2010), etc. Furthermore, there has been some effort in generalizing the Galileon to a non-flat background. The first attempt was then to covariantize directly the decoupling limit and to study its resulting cosmology (Chow and Khoury 2009). In particular, it was shown in Deffayet et al. (2009) that the naive covariantization would yield ghost-like terms at the level of equations of motion but a given unique nonminimal coupling between π and the curvature can remove these terms resulting in second order equations of motion (Deffayet et al. 2009), which are also consistent with a higher-dimensional construction (de Rham and Tolley 2010). While this covariantization is ghost-free, the Galileon symmetry is broken explicitly in curved backgrounds. However, there has been a successful generalization to the maximally symmetric backgrounds (Burrage et al. 2011; Goon et al. 2011).

There exists a parallel to theories centered on a massive graviton: Galileon-type interaction terms naturally arise in gravitational theories using a massive spin-2 particle as an exchange particle, which has, in addition, been constructed to be ghost-free, be it in three dimensions, Bergshoeff et al. (2009), de Rham et al. (2011) or for a generalized Fierz-Pauli action in four dimensions (de Rham et al. 2011; de Rham and Gabadadze 2010b). Such a theory was also constructed using auxiliary extra dimensions, de Rham and Gabadadze (2010a). The massive graviton of spin 2, has a total of 5 degrees of freedom. These degrees of freedom show different behaviours in the decoupling limit, namely 2 helicity-2 modes, 2 helicity-1 modes which decouple from the other degrees of freedom, and one helicity-0 mode, which again does not decouple giving rise to the vDVZ-discontinuity. As in the braneworld models presented previously, this can be cured by invoking the Vainshtein-mechanism in which the scalar mode appears as a scalar field with second order derivative interaction terms in the equation of motion. Not only is the existence of a graviton mass a fundamental question from a theoretical perspective, it could also have important consequences both in cosmology and in solar system physics, Koyama et al. (2011), Chkareuli and Pirtskhalava (2012). Although solar system observations have confirmed General Relativity to high accuracy and placed bounds on the graviton mass to be smaller than a few $\sim 10^{-32}$ eV, even such a small mass would become relevant at the Hubble scale which corresponds to the graviton Compton wavelength.

While the self-accelerating solutions in the above models yield viable expansion histories including late-time acceleration, they do not address the cosmological constant problem, i.e., the giant mismatch between the theoretically computed high energy density of the vacuum and the low observed value. A possible answer comes from the idea of degravitation, which asserts that the energy density could be as large as the theoretically expected value, but would not bear a large effect on the geometry. Technically, gravity is less strong on large scales (IR-limit) and could act as a high-pass filter suppressing the gravitational effect of a potentially large vacuum energy. Since such modifications of gravity in the IR naturally arise in models of massive gravity, they logically provide a possible mechanism to degravitate the vacuum energy density, Dvali et al. (2002, 2003, 2007), Arkani-Hamed et al. (2002). Analogously, the DGP braneworld model can be extended to higher dimensions to tackle the cosmological constant problem as well, Dvali et al. (2007), Gabadadze and Shifman (2004), de Rham et al. (2008).

All these models of infrared modifications of general relativity are united by the common feature of invoking new degrees of freedom. These degrees of freedom in the considered space-time are particles characterized by their masses and spins (or equivalently helicities). These particles are the excited quanta of the underlying fields. Their Lagrangian are constructed based on the requirement of yielding second order equations of motion, hence with a bounded Hamiltonian from below. Let us briefly discuss the zoo of these new degrees of freedom and their property based on their masses, spins and interactions given by the symmetries.

1.1.2 Quantum Field Theories

In the standard model of particle physics as well as in cosmology the mostly studied fields are those with particles (massless or massive) of spin-0, 1/2, 1 and 2. Particles with higher spin only arise in theories beyond the standard model. Particles with zero mass can be described by their helicities, i.e. how they change under rotational transformations transversal to the motion direction. For long range forces only massless bosons (or with very light masses) come into consideration since forces carried by massive particles have the exponential Yukawa suppression. In this section, we will introduce the protagonists of the particle zoo and discuss their main properties.

1.1.2.1 Spin-0 Fields

After a tremendous effort, physicists have finally succeeded in finding the missing piece of the Standard Model of particle physics at Cern: the Higgs boson. This constitutes the so far only observed fundamental spin-0 particle in nature. This outstanding experimental discovery motivates now more than ever the study of scalar degrees of freedom as a candidate for dark energy since we now know that they really exist in nature. Nevertheless, this is not the only reason why they are by far the most extensively explored candidates in cosmology. On an equal footing of importance the other reason for considering scalar fields is that they can provide accelerated expansion without breaking the isotropy of the universe. In addition to this practical reason, scalar fields arise in a very natural manner in modified theories of gravity or high energy physics. Demanding Lorentz invariance, the Lagrangian for a scalar field π, with the consistent self-interactions can be constructed very easily:

$$\mathcal{L}_\pi = -\frac{1}{2}(\partial\pi)^2 - V(\pi) \tag{1.3}$$

The construction of a mass term for the scalar field is trivial $\frac{1}{2}m_\pi^2\pi^2$ and is contained in the general potential term. In fact, the most general renormalizable potential for a scalar field in 4 dimensions only contains up to quartic powers of the scalar field. Also notice that adding a mass term (or a potential in general) for the scalar field does not alter the number of propagating physical degrees of freedom, since there is no gauge symmetry to be broken.

As a candidate for dark energy, one obstruction that one usually meets is that the effective mass of the scalar field must be very small (of the order of today's Hubble constant $H_0 \simeq 10^{-33}$ eV). Thus, if this light scalar degree of freedom couples to ordinary matter, then it can mediate a fifth force with a long range of interaction which has never been detected in Solar System gravity tests or laboratory experiments. On that account, it is crucial to reconcile the existence of a current phase of accelerated expansion driven by a light scalar field on very large scales with the absence of fifth forces on small scales. One could fine-tune its coupling to matter

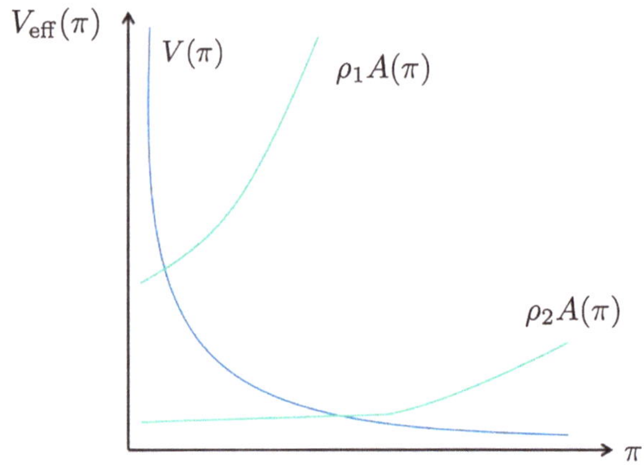

Fig. 1.2 The effective potential of the chameleon scalar field π

which is less satisfactory. Fortunately, there exist alternatives to fine-tunings thanks to the screening mechanisms that allow to hide the scalar field on small scales while being unleashed on large scales to produce cosmological effects. Typical examples of screening mechanisms are Vainshtein, chameleon or symmetron.[2] In the chameleon mechanism the important ingredient is a conformal coupling between the scalar and the matter fields $\mathcal{L}_{\text{matter}}[\tilde{g}_{\mu\nu} = g_{\mu\nu}A^2(\pi)]$ such that the equation of motion for a static configuration of the scalar field becomes (Khoury and Weltman 2004)

$$\nabla^2\pi = V_{,\pi} - A^3 A_{,\pi}\tilde{T} = V_{,\pi} + A_{,\pi}\rho \tag{1.4}$$

where $\tilde{T} \sim \rho/A^3$. As it is clear from the equation of motion, the conformal coupling to the matter fields gives rise to an effective potential which depends explicitly on the environmental density:

$$V_{\text{eff}}(\pi) = V(\pi) + \rho A(\pi). \tag{1.5}$$

This means that the mass of this new degree of freedom as a scalar field depends on the local density

$$m^2_{\text{min}}(\pi) = V_{,\pi\pi}(\pi_{\text{min}}) + \rho A_{,\pi\pi}(\pi_{\text{min}}). \tag{1.6}$$

Depending on the choice of the potential $V(\pi)$ and the conformal coupling $A(\pi)$, the mass of the scalar field can be made large in regions of high density and so screen the scalar field (Fig. 1.2).

The symmetron screening mechanism is conceptually very similar to the Chameleon mechanism even though the realization is slightly different. Again the important ingredient is a conformal coupling to the matter fields but with a very

[2]One can basically use the mass term, the coupling to matter or the kinetic term of the scalar field in order to achieve screening.

specific function $A(\pi)$ and potential (Hinterbichler and Khoury 2010)

$$\begin{aligned} A(\pi) &= 1 + \frac{\pi^2}{2M^2} + \cdots \\ V(\pi) &= -\frac{1}{2}\mu_1^2\pi^2 + \frac{1}{4}\mu_2\pi^4 \end{aligned} \tag{1.7}$$

with the free parameters μ_1, μ_2, M. Again the conformal coupling results in an effective potential of the form

$$V_{\text{eff}}(\pi) = \left(\frac{1}{2}\mu_1^2 - \frac{\rho}{2M^2}\right)\pi^2 + \frac{1}{4}\mu_2\pi^4. \tag{1.8}$$

The perturbations of the scalar field couple to the matter fields as $\frac{\bar{\pi}}{M^2}\delta\pi\rho$. In high density symmetry-restoring environments $\rho > \mu_1^2 M^2$, the scalar field sits in a minimum at the origin with the vacuum expectation value (VEV) ~ 0 and so the fluctuations of the field do not couple to matter. As the local density drops, the symmetry of the field is spontaneously broken and the field falls into one of the two new minima with a non-zero VEV. Hence, the coupling to matter depends on the environment, becoming small in regions of high density (Fig. 1.3).

Last but not least the Vainshtein mechanism relies on the strong derivative self-interactions of the scalar degree of freedom. At the classical level the background configuration relies on non-linearities being large $\Lambda_3^{-3}\pi(\partial\pi)^2 \gg 1$ but perturbations on top of this classical background configurations are weakly coupled. Consider a localized source $T = M\delta^{(3)}(r) + \delta T$ and perturbations of the scalar field $\pi = \bar{\pi}(r) + \delta\pi(x^\mu)$. The Vainshtein mechanism works by modifying the kinetic matrix symbolically as

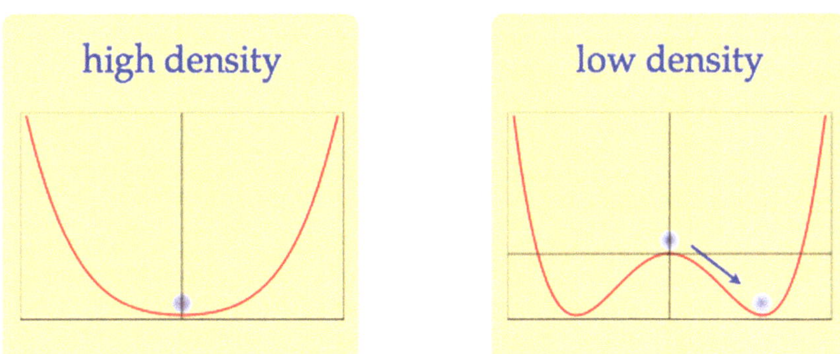

Fig. 1.3 The effective potential of the symmetron scalar field π in two different density regimes

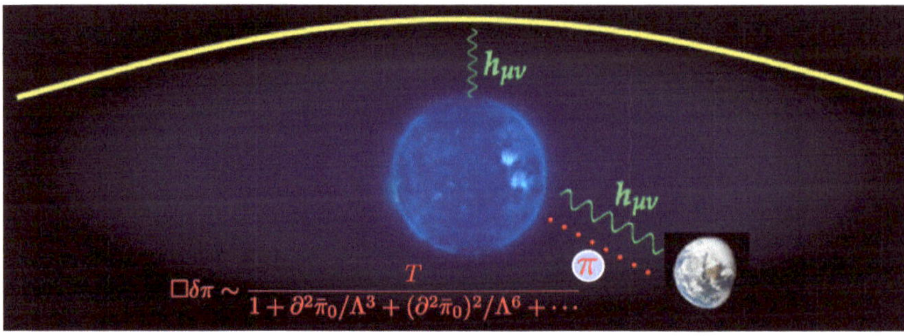

Fig. 1.4 Between two massive objects, there is not only the exchange of the helicity-2 field $h_{\mu\nu}$, but also of the helicity-0 degree of freedom π. However, the latter becomes screened due to derivative self-interactions

$$\mathcal{L} = -\frac{1}{2}\left(1+\frac{\partial^2\bar{\pi}_0}{\Lambda^3}+\frac{(\partial^2\bar{\pi}_0)^2}{\Lambda^6}+\cdots\right)(\partial\delta\pi)^2+\frac{1}{M_p}\delta\pi\delta T. \tag{1.9}$$

After properly canonically normalizing the field, the effective coupling to matter depends on the self-interactions of the scalar field (Vainshtein 1972; Deffayet et al. 2002)

$$\mathcal{L} = -\frac{1}{2}(\partial\delta\pi)^2+\frac{1}{M_p}\frac{\delta\pi\delta T}{\sqrt{(1+\frac{\partial^2\bar{\pi}_0}{\Lambda^3}+\frac{(\partial^2\bar{\pi}_0)^2}{\Lambda^6}+\cdots)}}. \tag{1.10}$$

The coupling to matter becomes small for strongly self-interacting fields $(1+\frac{\partial^2\bar{\pi}_0}{\Lambda^3}+\frac{(\partial^2\bar{\pi}_0)^2}{\Lambda^6}+\cdots)\gg 1$. As we mentioned before the strength with which this new scalar degree of freedom can couple to the standard model fields is highly constrained by searches for fifth forces and violations of the weak equivalence principle and is typically required to be orders of magnitude weaker than gravity. Thanks to these screening mechanisms the new scalar degrees of freedom can naturally couple to standard model fields, whilst still being in agreement with observations and source the acceleration of the universe (Fig. 1.4).

1.1.2.2 Spin-1/2 Fields

The Standard Model is rich in spin 1/2 particles. It comprises two important families of elementary fermions, the leptons and quarks. They obey the Pauli exclusion principle, meaning that only one fermion can occupy a quantum state at the same time. Fermions come in three different types, namely the massless Weyl fermions, the massive Dirac fermions and Majorana fermions. Nevertheless, most of the Standard Model fermions are Dirac fermions. We can describe the Dirac fermion by the following Lagrangian

$$\mathcal{L}_\psi = \bar{\psi}(i\gamma^\mu\partial_\mu - m)\psi \tag{1.11}$$

where ψ is the Dirac spinor and $\bar{\psi} \equiv \psi^{\dagger}\gamma^0$. The γ matrices generate the Clifford algebra $\{\gamma^\mu, \gamma^\nu\} = 2\eta^{\mu\nu}$ and, in the Dirac representation, are given in terms of the Pauli matrices σ^i. In the cosmological evolution, standard spinorial fields have been much less intensively explored than bosonic fields. One reason for this is the difficulty in interpreting classical fermionic fields in terms of their underlying quantum particles. Since fermions cannot condensate in coherent states, they cannot produce classical fermionic fields. Of course, fermions can play a relevant role in the cosmological evolution as a thermal distribution, as it happens for instance with neutrinos. The second reason is related to the fast decay of fermions in an expanding universe and the inefficient production of fermions during preheating.

1.1.2.3 Spin-1 Fields

The Standard Model of elementary particles contains both abelian (photon) and non-abelian vector fields (weak and strong interactions carriers) as the fundamental fields of the gauge interactions. They come in both as massless and massive vector fields. Therefore this motivates an exploration of the role of vector fields (not necessarily those of the standard model) in the cosmological evolution. Vector fields also arise in a natural manner in modified theories of gravity or high energy physics. Nevertheless, vector fields in cosmology have the additional difficulty with respect to scalars that they naturally lead to the presence of large scale anisotropic expansion that could conflict the high isotropy observed in the CMB. However, they could be used to explain the reported anomalies by WMAP and Planck in cosmological observations at large scales that could be signalling the presence of a preferred direction in the universe. The Lagrangian for a massless vector field must be constructed such that it is invariant under the gauge symmetry

$$A_\mu \rightarrow A_\mu + \partial_\mu \theta. \tag{1.12}$$

The gauge symmetry is mandatory in order to have two propagating degrees of freedom. This requirement uniquely leads to Maxwell theory

$$\mathcal{L}_{A_\mu} = -\frac{1}{4}F_{\mu\nu}^2 - J_\mu A^\mu, \tag{1.13}$$

where $F_{\mu\nu} = \partial_\mu A_\nu - \partial_\nu A_\mu$ is the field strength and J_μ is an external source. The equations of motion are simply given by

$$\partial_\nu F^{\mu\nu} = J^\mu. \tag{1.14}$$

Taking the divergence of the equation of motion yields $\partial_\mu J^\mu = 0$ and hence the external source must be conserved. Since we have the gauge symmetry, we can choose a gauge, for instance the Lorenz condition $\partial_\mu A^\mu = 0$. This gauge choice brings the equations of motion into the form $\Box A^\mu = J^\mu$. Together with the residual gauge $\Box\theta = 0$ this kills the two unphysical modes.

We can add a mass term to the Maxwell action by explicitly breaking the gauge symmetry

$$\mathcal{L}_{A_\mu} = -\frac{1}{4}F_{\mu\nu}^2 - \frac{1}{2}m_A^2 A_\mu^2 - J_\mu A^\mu \tag{1.15}$$

yielding a massive spin-1 theory with three propagating degrees of freedom (see Sect. 3.7 in Chap. 3 for a more general Lagrangian with derivative interactions yielding three propagating degrees of freedom). The equations of motion change to

$$\partial_\nu F^{\mu\nu} - m_A^2 A^\mu = J^\mu \tag{1.16}$$

Now taking the divergence of the equations of motion gives the constraint

$$- m_A^2 \partial_\mu A^\mu = \partial_\mu J^\mu \tag{1.17}$$

For a conserved current the equation of motion becomes simply a Klein-Gordon equation $(\Box - m_A^2)A^\mu = J^\mu$, together with the condition $\partial_\mu A^\mu = 0$.

We can restore the gauge invariance using the Stueckelberg trick. For this we add an additional scalar field via

$$A_\mu \to A_\mu + \partial_\mu \pi \tag{1.18}$$

such that the action for the massive spin-1 field becomes

$$\mathcal{L}_{A_\mu} = -\frac{1}{4}F_{\mu\nu}^2 - \frac{1}{2}m_A^2 (A_\mu + \partial_\mu \pi)^2 - J_\mu (A^\mu + \partial_\mu \pi) \tag{1.19}$$

making the action now again invariant under the simultaneous transformations $A_\mu \to A_\mu + \partial_\mu \theta$ and $\pi \to \pi - \theta$. After canonically normalizing the additional field $\pi \to \frac{1}{m_A}\pi$ the interactions can be expressed as (Fierz and Pauli 1939)

$$\mathcal{L}_{A_\mu} = -\frac{1}{4}F_{\mu\nu}^2 - \frac{1}{2}m_A^2 A_\mu^2 - \frac{1}{2}(\partial\pi)^2 - m_A A_\mu \partial^\mu \pi - J_\mu \left(A_\mu + \frac{\partial_\mu \pi}{m_A}\right) \tag{1.20}$$

Now taking the $m_A \to 0$ limit for a conserved source results in a theory of a massless scalar field completely decoupled from a massless vector field

$$\mathcal{L}_{A_\mu} = -\frac{1}{4}F_{\mu\nu}^2 - \frac{1}{2}\partial\pi^2 - J_\mu A^\mu \tag{1.21}$$

This is the reason why taking the $m_A \to 0$ limit does not give rise to the vDvZ discontinuity in the case of massive vector fields.

Vector fields have extensively been investigated in cosmological scenarious as candidates to explain the current phase of accelerated expansion (Boehmer and Harko 2007) or to drive the inflationary epoch (Golovnev et al. 2008) or to generate magnetic fields during inflation using non-minimal couplings (Turner and Widrow 1988). There has been also some attempts to screen the vector field on small scales (Jimenez et al. 2013).

1.1.2.4 Spin-2 Field

Similarly as in the massless spin-1 case, the theory for a pure massless spin-2 field needs to have a gauge symmetry in order to have two propagating degrees of freedom. This uniquely leads to general relativity with the action

$$S = \int d^4x\, (\mathcal{L}_{\rm GR} + \mathcal{L}_{\rm matter}) \tag{1.22}$$

$$= \frac{M_{\rm Pl}^2}{2}\int d^4x\sqrt{-g}R + \int d^4x\mathcal{L}_{\rm matter}\,. \tag{1.23}$$

This Lagrangian is invariant under full general coordinate transformations which in the linearized limit corresponds to the invariance under the gauge symmetry

$$h_{\mu\nu} \to h_{\mu\nu} + \partial_\mu \xi_\nu + \partial_\nu \xi_\mu \tag{1.24}$$

once one expands the action to second order in the metric perturbations around flat space-time

$$g_{\mu\nu} = \eta_{\mu\nu} + \frac{2}{M_{\rm Pl}}h_{\mu\nu} \quad \text{and} \quad g^{\mu\nu} = \eta^{\mu\nu} - \frac{2}{M_{\rm p}}h^{\mu\nu} + \frac{4}{M_{\rm p}^2}h^{\mu\alpha}h^{\nu}_{\alpha} + \cdots\,. \tag{1.25}$$

The full Lagrangian to second order in h is

$$\mathcal{L} = -h^{\mu\nu}\hat{\mathcal{E}}^{\alpha\beta}_{\mu\nu}h_{\alpha\beta} + \frac{1}{M_{\rm Pl}}h_{\mu\nu}T^{\mu\nu} + \frac{1}{2M_{\rm Pl}^2}h_{\mu\nu}h_{\alpha\beta}T^{\mu\nu\alpha\beta}\,, \tag{1.26}$$

where $\hat{\mathcal{E}}$ is the Lichnerowicz operator

$$\hat{\mathcal{E}}^{\alpha\beta}_{\mu\nu}h_{\alpha\beta} = -\frac{1}{2}\left(\Box h_{\mu\nu} - 2\partial_\alpha\partial_{(\mu}h^{\alpha}_{\nu)} + \partial_\mu\partial_\nu h - \eta_{\mu\nu}(\Box h - \partial_\alpha\partial_\beta h^{\alpha\beta})\right)\,, \tag{1.27}$$

and $T_{\mu\nu}$ is the stress-energy tensor, whilst $T^{\mu\nu\alpha\beta}$ is its derivative with respect to the metric,

$$T_{\mu\nu} = \frac{-2}{\sqrt{-g}}\frac{\delta\sqrt{-g}\mathcal{L}_m}{\delta g^{\mu\nu}} \quad \text{and} \quad T_{\mu\nu\alpha\beta} = -\frac{2}{\sqrt{-g}}\frac{\delta\sqrt{-g}T_{\mu\nu}}{\delta g^{\alpha\beta}} \tag{1.28}$$

At first order in perturbation the Einstein equations simplify to

$$\hat{\mathcal{E}}_{\mu\nu}{}^{\alpha\beta}h_{\alpha\beta} = \frac{1}{2M_p}T_{\mu\nu}. \tag{1.29}$$

Taking the divergence of the equation of motion yields again the conservation of external sources $\partial_\mu T^{\mu\nu} = 0$. Choosing the Lorenz gauge $\partial^\mu \bar{h}_{\mu\nu} = 0$ the equations of motion simplify to $-\frac{1}{2}\Box\bar{h}_{\mu\nu} = \frac{1}{2M_p}T_{\mu\nu}$, where $\bar{h}_{\mu\nu} = h_{\mu\nu} - \frac{1}{2}\eta_{\mu\nu}h$. Together with the residual gauge symmetry $\Box\xi_\alpha = 0$ this gauge choice eliminates eight out of ten degrees of freedom.

The question whether or not the graviton is massless is a fundamental question from a theoretical perspective. Is the graviton really massless or does its mass just happen to be so small that it can be safely neglected on sufficiently small distance scales? How do you make the graviton massive? You might naively start with an analog ansatz to the case of Proca field $m^2 g^{\mu\nu}A_\mu A_\nu$ by writing the massive gravity as

$$\sqrt{-g}\left[\frac{m^2}{2}\left(c_1 g_{\mu\nu}g^{\mu\nu} + c_2 g^\mu_\mu g^\nu_\nu\right)\right] \tag{1.30}$$

but very soon you realize that this just corresponds to a cosmological constant rather than a mass term. You could then try with the ansatz m^2R^2 and also very soon realize that this contains derivatives of $g_{\mu\nu}$ and can not be a valid mass term. Sooner or later you would end up with the more promising ansatz

$$\frac{m^2}{2}\left(c_1 h_{\mu\nu}h^{\mu\nu} + c_2 h^\mu_\mu h^\nu_\nu\right) \tag{1.31}$$

once the metric fluctuations are expanded around flat space-time $g_{\mu\nu} = \eta_{\mu\nu} + \frac{2}{M_{Pl}}h_{\mu\nu}$. Fierz-Pauli was the first successful construction of a linear mass term without giving rise to any Boulware-Deser ghost degree of freedom with the restriction $c_2 = -c_1$. One can show that away from the Fierz-Pauli tuning the mass of the ghost degree of freedom would correspond to $m_g^2 = -\frac{c_1(c_1+4c_2)m^2}{4(c_1+c_2)}$ which goes to infinity for the Fierz-Pauli tuning. Thus, the safe linear mass term for the graviton reads

$$\mathcal{L} = -h^{\mu\nu}\hat{\mathcal{E}}^{\alpha\beta}_{\mu\nu}h_{\alpha\beta} - \frac{m^2}{2}\left(h_{\mu\nu}h^{\mu\nu} - h^\mu_\mu h^\nu_\nu\right) + \frac{1}{M_{Pl}}h_{\mu\nu}T^{\mu\nu}. \tag{1.32}$$

The equations of motion yield

$$-2\hat{\mathcal{E}}_{\mu\nu}{}^{\alpha\beta}h_{\alpha\beta} - m^2\left(h_{\mu\nu} - \eta_{\mu\nu}h\right) = -\frac{1}{M_p}T_{\mu\nu} \tag{1.33}$$

Taking the divergence of the equations of motion implies $m^2(\partial^\mu h_{\mu\nu} - \partial_\nu h) = \frac{1}{M_p}\partial^\mu T_{\mu\nu}$. In the case of a conserved source this is simply the statement that $\partial^\mu h_{\mu\nu} =$

$\partial_\nu h$. Furthermore, taking the trace of the equation of motion gives rise to $-3m^2 h = \frac{T}{M_p}$. These constraints equations allow us to get rid of the five unphysical degrees of freedom. Plugging these relations back into the field equations results in

$$(\Box - m^2)h_{\mu\nu} = -\frac{1}{M_p}\left(T_{\mu\nu} - \frac{1}{3}\eta_{\mu\nu}T - \frac{1}{3m^2}\partial_\mu\partial_\nu T\right) \tag{1.34}$$

Because of the mass term one has lost the invariance under general coordinate transformations. However, the diffeomorphism invariance can be restored by introducing the Stueckelberg fields in a similar way as for the massive spin-1 field. This can be achieved by defining new fields in the form (Siegel 1986)

$$h_{\mu\nu} \rightarrow h_{\mu\nu} + \partial_\mu A_\nu + \partial_\nu A_\mu \quad \text{and} \quad A_\mu \rightarrow A_\mu + \partial_\mu \pi \tag{1.35}$$

which once plugged in back into the lagrangian gives

$$\begin{aligned}\mathcal{L} = -h^{\mu\nu}\hat{\mathcal{E}}^{\alpha\beta}_{\mu\nu}h_{\alpha\beta} - \frac{m^2}{2}\left(h_{\mu\nu}h^{\mu\nu} - h^\mu_\mu h^\nu_\nu\right) + \frac{1}{M_{Pl}}h_{\mu\nu}T^{\mu\nu} \\ -\frac{m^2}{2}F^2_{\mu\nu} - 2m^2(h_{\mu\nu}\partial^\mu A^\nu - h\partial A) - \frac{2}{M_{Pl}}A_\mu\partial_\nu T^{\mu\nu} \\ -2m^2(h_{\mu\nu}\partial^\mu\partial^\nu\pi - h\partial^2\pi) + \frac{2}{M_{Pl}}\pi\partial_\mu\partial_\nu T^{\mu\nu}\end{aligned} \tag{1.36}$$

This lagrangian is now invariant under the transformations

$$\begin{aligned} h_{\mu\nu} \rightarrow h_{\mu\nu} + \partial_\mu\xi_\nu + \partial_\nu\xi_\mu \quad &\text{and} \quad A_\mu \rightarrow A_\mu - \xi_\mu \\ A_\mu \rightarrow A_\mu + \partial_\mu\theta \quad &\text{and} \quad \pi \rightarrow \pi - \theta \end{aligned} \tag{1.37}$$

After canonically normalizing the fields $A_\mu \rightarrow \frac{1}{m}A_\mu$ and $\pi \rightarrow \frac{1}{m^2}\pi$ and taking the $m \rightarrow 0$ limit, the Lagrangian simply becomes (for conserved sources)

$$\begin{aligned}\mathcal{L} = -h^{\mu\nu}\hat{\mathcal{E}}^{\alpha\beta}_{\mu\nu}h_{\alpha\beta} - \frac{1}{2}F^2_{\mu\nu} + \frac{1}{M_{Pl}}h_{\mu\nu}T^{\mu\nu} \\ -2(h_{\mu\nu}\partial^\mu\partial^\nu\pi - h\partial^2\pi)\end{aligned} \tag{1.38}$$

This corresponds to a scalar-vector-tensor theory in which the scalar field is kinetically mixed with the tensor whereas the vector field is completely decoupled.

In the next Sect. 1.2 we will try to generalize the interactions of the spin-2 field to the non-linear case and discuss the difficulties one usually encounters. But before doing that let us first introduce the concept of Effective Field Theories since it became an essential tool in particle physics and modern cosmology.

1.1.3 Effective Field Theories

Nature discloses itself at different energy scales. Indeed the atomic physics scale and the galactic physics scale are the two edges of a large hierarchy of scales. In order to study a given physical system it is an indispensable task to find a framework in which the most relevant physics at that scale are captured by a simple description without having to understand everything from the rest. This means that one can study the low-energy physics, independently of the specific aspects of the high-energy physics. The appropriate framework is the framework of effective field theory. It essentially describes the low energy physics below the energy scale Λ_c after integrating out all the degrees of freedom beyond Λ_c. The resulting low-energy Lagrangian is an expansion in powers of $1/\Lambda_c$ of the non-renormalizable interactions among the light degrees of freedom incorporating all important symmetries of the underlying fundamental theory. In order to construct the effective field theory of the physical problem one needs to

- specify the fields content and the underlying symmetries of the physical problem
- write down the most general Lagrangian containing all terms allowed by the symmetry.

According to the effective field theory approach, the terms in the Lagrangian can be classified into relevant, marginal and irrelevant operators of dimensions $d < 4$, $d = 4$ and $d > 4$ respectively. The relevant and marginal operators can be renormalized while the irrelevant operators are not renormalizable. At low energies, the contribution of the non-renormalizable irrelevant operators are negligible since they come in as inversely proportional powers of the cut-off scale Λ_c. On the other hand, the relevant operators give rise to contributions with positive powers of the cut-off. An effective field theory with only marginal operators is a good renormalizable low energy effective field theory. With the techniques of effective field theory we are now at a better position to understand the physical interpretation of non-renormalizable theories. Even if a theory can not be renormalized, its quantization can make perfect sense assuming we are applying it only to the low-energy physics.

Let us concretize this by looking at a few explicit examples. Our Lagrangian for the spin-0 field in Eq. 1.3 can be extended to

$$\begin{aligned}\mathcal{L}_\pi = &-\frac{1}{2}(\partial\pi)^2 - \frac{1}{2}m_\pi^2\pi^2 + c_1\pi^4 + \frac{c_2}{\mu_1^2}\pi^6 + \frac{c_3}{\mu_2^4}\pi^8 + \cdots \\ &+\frac{d_1}{\mu_3^2}\pi^2(\partial\pi)^2 + \frac{d_2}{\mu_4^4}\pi^4(\partial\pi)^2 + \cdots \end{aligned} \tag{1.39}$$

This Lagrangian can be considered as an effective field theory describing the dynamics of the light scalar degree of freedom π which contains all the important symmetries of the underlying fundamental theory. The requirement of the symmetry of the system and locality constraints these interactions. As an effective field theory, the operators $(\partial\pi)^2$ and π^2 with dimension $d < 4$ correspond to the relevant operators, the operator π^4 with dimension $d = 4$ to a marginal operator and all the remaining operators with dimension $d > 4$ are the irrelevant operators. The cutoff scale is the lowest among the scales $\mu_1, \mu_2, \mu_3 \cdots$. And once the energies considered go beyond this cutoff scale, an infinite number of interactions need to be taken into account. This Lagrangian can be further restrained to a smaller subgroup of interactions by demanding additional symmetries like shift symmetry or Galileon symmetry.

As a next example consider the light by light scattering in Quantum Electrodynamics (QED) at very low photon energies. The QED Lagrangian is given by

$$\mathcal{L}_{\mathrm{QED}} = -\frac{1}{4}F_{\mu\nu}^2 + \bar{\psi}(i\gamma^\mu\partial_\mu - m_\psi)\psi - e\bar{\psi}\gamma^\mu A_\mu\psi \tag{1.40}$$

In the low-energy regime, where the photon energies are much smaller than m_ψ, the physics can be rather described by the effective Lagrangian

$$\mathcal{L}_{\mathrm{QED}} = -\frac{1}{4}F_{\mu\nu}^2 + \frac{c_1}{m_\psi^4}(F_{\mu\nu}F^{\mu\nu})^2 + \frac{c_2}{m_\psi^4}(F_{\mu\nu}\tilde{F}^{\mu\nu})^2 + \cdots \tag{1.41}$$

where $\tilde{F}_{\mu\nu} = \epsilon_{\mu\nu\alpha\beta}F^{\alpha\beta}$ is the dual field strength tensor. This Lagrangian can be obtained by integrating out the electron field Ψ from the original QED generating functional or equivalently by calculating the lowest order light-by-light box diagram given by a single electron loop. However, one would also obtain this Lagrangian as a consequence of the imposed symmetries of gauge, lorentz, charge conjugation and parity invariance. The scaling of the constants can then be estimated through a naive power counting (Burgess 2007).

Another very illustrative example comes from Quantum Chromodynamics (QCD). The Lagrangian of QCD is given by

$$\mathcal{L}_{\mathrm{QCD}} = -\frac{1}{2}G_{\mu\nu}^2 + \bar{q}(i\gamma^\mu\partial_\mu - m_q)q + g\bar{q}\gamma^\mu G_\mu q \tag{1.42}$$

where q represents the quark and $G_{\mu\nu}$ the gluon field strength tensor

$$G_{\mu\nu} = \partial_\mu G_\nu - \partial_\nu G_\mu - ig[G_\mu, G_\nu] \tag{1.43}$$

with G_μ being the gluon field. The quarks differ in their masses. One can divide them into two groups, the low mass quarks: up, down, strange quarks and the heavy quarks: charm, bottom, top quarks. One can now construct the effective field theory in the low energy limit in which the c, b, t quarks can be treated as infinitely heavy whereas the light mass quarks can be approximated as massless quarks. In this chiral limit the QCD Lagrangian becomes

$$\mathcal{L}_{\mathrm{QCD}} = -\frac{1}{2}G_{\mu\nu}^2 + \bar{q}i\gamma^\mu\partial_\mu q + g\bar{q}\gamma^\mu G_\mu q \tag{1.44}$$

with the extra chiral symmetry (Scherer 2003). This is the essence of the Chiral Perturbation Theory which is used in order to study hadronic physics.

We have been giving examples from QED and QCD and actually General Relativity itself is also an effective field theory valid at low energies below M_{Pl}. General Relativity can be perfectly quantized in the effective field theory sense provided that we are working within a regime where the irrelevant operators are small compared to the relevant ones. Loop calculations give higher order derivative interactions of the form (Burgess 2007)

$$\mathcal{L}_{GR} = \sqrt{-g}\left(\frac{M_{Pl}^2}{2}R + c_1R^2 + c_2R_{\mu\nu}R^{\mu\nu} + c_3R_{\mu\nu\alpha\beta}R^{\mu\nu\alpha\beta} + \frac{d_1}{M_{Pl}^2}R^3 + \cdots\right) \tag{1.45}$$

which would need to be taken into account above energies beyond M_{Pl}, meaning that these irrelevant operators start dominating once we go beyond the cut off scale of the effective field theory.

Throughout this thesis, we shall be working with non-renormalizable theories and interactions. As discussed in this section, this does not necessarily represent a flaw of the theories, but rather they should be regarded as effective field theories and treated as such. At this respect, it is a crucial aspect for their reliability to compute the corresponding cut-off scale below which the theory is sensible. This is not a straightforward computation and some debate about the true cut-off scale of a given theory still exists in the literature.

1.2 Massive Gravity

Brimming over with enthusiasm for having found the linear ghost free massive spin-2 field (1.32) one can take a step forward and compute the graviton exchange amplitude between two sources. What one rather encounters is a worrying result. In the limit of vanishing graviton mass one does not recover the general relativity result for the exchange amplitude between the sources. Actually this result is not surprising at all. After doing the field redefinition $h_{\mu\nu} = \tilde{h}_{\mu\nu} + \pi\eta_{\mu\nu}$ the mixing between the scalar and tensor interactions in Eq. (1.38) vanish at the expense of the coupling of the scalar field to the stress energy tensor πT. It is exactly this coupling that gives rise to the vDVZ discontinuity which survives the $m \to 0$ limit. However, we were working in a regime in which some of the degrees of freedom of the massive graviton were interacting highly non-linearly and therefore the vDVZ discontinuity is just an artifact of the linear approximation. Unfortunately introducing a non-linear mass term for the graviton is not as easy as it might look at first sight.[3] The ghost we have cured by Fierz-Pauli tuning comes back at non-linear level (the sixth degree of freedom is

[3]The non-linear extension for the spin-1 field is trivial in the sense that one does not have to deal with the Boulware-Deser ghost.

associated to higher derivative terms for the helicity-0 degree of freedom). For the construction of a mass term the requirements are that it can not contain derivative interactions for $g_{\mu\nu}$ and that it should only contain five physical degrees of freedom with a positive kinetic energy. The only quantities without derivative interactions are det$g$ corresponding to a cosmological constant and tr$g$. Therefore one is forced to introduce an extra auxiliary rank two metric $f_{\mu\nu}$ on top of which the massive graviton can propagate. The de Rham-Gabadadze-Tolley (dRGT) model of massive gravity resolves the ghost-problem, being the first example of a ghost free non-linear theory of massive gravity in arbitrary dimensions. In this original constructions the reference metric was chosen to be flat $f_{\mu\nu} = \eta_{\mu\nu}$. In general the mass term can be constructed out of the scalar functions of $g^{\mu\alpha}f_{\alpha\nu}$. As we mentioned in the previous chapter the action (1.32) does not have the invariance under general coordinate transformations and either $\eta_{\mu\nu}$ nor $h_{\mu\nu}$ are real tensors. Instead, for the construction of the higher non-linear interactions we shall work in a framework in which the diffeomorphism invariance is restored by introducing the Stueckelberg fields. For this purpose lets define a tensor $H_{\mu\nu}$ of the following form

$$H_{\mu\nu} = g_{\mu\nu} - \eta_{ab}\partial_\mu\varphi^a\partial_\nu\varphi^b \tag{1.46}$$

which will enter in the scalar functions of the mass term $g^{\mu\alpha}f_{\alpha\nu} = \delta^\mu_\nu - H^\mu_\nu$. The tensor $H_{\mu\nu}$ is now a covariant tensor as long as the four fields φ^a transform as scalars under a change of coordinates (Arkani-Hamed et al. 2003). Next step consists of writing down all the possible contractions for the tensor $H_{\mu\nu}$ to construct the most general potential (de Rham and Gabadadze 2010b)

$$\mathcal{L} = M_{Pl}^2\sqrt{-g}R - \frac{M_{Pl}^2 m^2}{4}\sqrt{-g}\,(U_2(g,H) + U_3(g,H) + U_4(g,H)\cdots) \tag{1.47}$$

with U_i's standing for the mass and potential terms of ith order in $H_{\mu\nu}$

$$\begin{aligned}
U_2(g,H) &= H_{\mu\nu}^2 - H^2\,,\\
U_3(g,H) &= c_1H_{\mu\nu}^3 + c_2HH_{\mu\nu}^2 + c_3H^3\,,\\
U_4(g,H) &= d_1H_{\mu\nu}^4 + d_2HH_{\mu\nu}^3 + d_3H_{\mu\nu}^2H_{\alpha\beta}^2 + d_4H^2H_{\mu\nu}^2 + d_5H^4\,,\\
U_5(g,H) &= f_1H_{\mu\nu}^5 + f_2HH_{\mu\nu}^4 + f_3H^2H_{\mu\nu}^3 + f_4H_{\alpha\beta}^2H_{\mu\nu}^3\\
&\quad + f_5H(H_{\mu\nu}^2)^2 + f_6H^3H_{\mu\nu}^2 + f_7H^5\,.
\end{aligned} \tag{1.48}$$

The coefficients c_i, d_i and f_i are a priori arbitrary. The four fields φ^a in $H_{\mu\nu}$ can be expanded in terms of the coordinates x^α and the fields π^α, as $\varphi^a = (x^\alpha - \pi^\alpha)\,\delta^a_\alpha$ (Arkani-Hamed et al. 2003) and we also expand the metric around Minkowski in the usual convention, $g_{\mu\nu} = \eta_{\mu\nu} + h_{\mu\nu}/M_{Pl}$, such that we obtain

$$H_{\mu\nu} = \frac{h_{\mu\nu}}{M_{Pl}} + \partial_\mu\pi_\nu + \partial_\nu\pi_\mu - \eta_{\alpha\beta}\partial_\mu\pi^\alpha\partial_\nu\pi^\beta\,. \tag{1.49}$$

The π_α's represent the Stückelberg fields that transform under reparametrization to guarantee that the tensor $H_{\mu\nu}$ in (1.49) transforms covariantly. In other words the covariant tensor $H_{\mu\nu}$ is a gauge transformed version of $h_{\mu\nu}$. In particular under linearized diffeomorphism, $x^\mu \to x^\mu + \frac{\xi^\mu}{M_{Pl}}$, the metric perturbations and the Stückelberg transform respectively as

$$h_{\mu\nu} \to h_{\mu\nu} - \partial_{(\mu}\,\xi_{\nu)} \qquad \text{and} \qquad \pi^\mu \to \pi^\mu + \frac{\xi^\mu}{M_{Pl}}\,. \tag{1.50}$$

In the unitary gauge where $\pi_\alpha = 0$ (or, $\varphi^a = x^\alpha \delta^a_\alpha$), Eq. (1.47) reduces to the standard Fierz-Pauli theory extended by a potential for the field $h_{\mu\nu}$. However, this is not always a convenient way of dealing with these degrees of freedom. For most of the studies in this thesis we will retain π_α and fix a gauge for $h_{\mu\nu}$ unless otherwise pointed out.

The theory (1.47) was studied in detail in de Rham and Gabadadze (2010a, b), and a two-parameter family of the coefficients was identified for which no sixth (Boulware-Deser ghost) degree of freedom arises. Interestingly, also extensions of GR by an extra auxiliary dimension (Gabadadze 2009; de Rham 2010), automatically generates the coefficients from this family at least up to the cubic order. In these theories the higher derivative nonlinear terms either cancel out, or organize themselves into total derivatives. Not only for a better intuition but for most of the purposes one can gain a quick insight into the physical properties of the theory by studying a specific limit, the so called decoupling limit. This limit creates a framework in which the most important physical properties of the theory become visible. For instance, the realization of the Vainshtein mechanism is best seen in the decoupling limit or if there is a Boulware-Deser ghost present in the theory, it can already be observed within the decoupling limit. Hence, the decoupling limit offers powerful tools to study the physical properties of the considered theory and we expect to recover analog results beyond this limit. For this ghost free theory of massive gravity, the decoupling limit is defined as follows

$$m \to 0, \quad M_{Pl} \to \infty, \quad \Lambda_3 = (M_{Pl} m^2)^{1/3} \ \text{fixed}. \tag{1.51}$$

In the decoupling limit we can ignore the helicity-1 modes as they do not couple to a conserved stress-tensor at the linearized level and focus on the helicity-2 and helicity-0 modes. At this point, it is worth pointing out that in the decoupling limit $M_{Pl} \to \infty$, the Stückelberg field π^μ ends up being gauge invariant under linearized diffeomorphism as clearly seen in Eq. (1.50).

We use the following decomposition for $H_{\mu\nu}$ in terms of the canonically normalized helicity-2 and helicity-0 fields after setting $\pi_a = \partial_a \pi/\Lambda_3^3$[4]

$$H_{\mu\nu} = \frac{h_{\mu\nu}}{M_{Pl}} + \frac{2\partial_\mu \partial_\nu \pi}{\Lambda_3^3} - \frac{\partial_\mu \partial^\alpha \pi \partial_\nu \partial_\alpha \pi}{\Lambda_3^6}. \tag{1.52}$$

The Lagrangian (1.47) with the two-parameter family of the coefficients reduces in the decoupling limit to the following expression (de Rham and Gabadadze 2010b)

$$\mathcal{L} = -\frac{1}{2} h^{\mu\nu} \mathcal{E}_{\mu\nu}^{\alpha\beta} h_{\alpha\beta} + h^{\mu\nu} \sum_{n=1}^{3} \frac{a_n}{\Lambda_3^{3(n-1)}} X_{\mu\nu}^{(n)}[\Pi], \tag{1.53}$$

where the first term represents the usual kinetic term for the graviton defined in the standard way with the Lichnerowicz operator given by (1.27), whereas $a_1 = -1/2$, and $a_{2,3}$ are two arbitrary constants, related to the two parameters from the set $\{c_i, d_i\}$ which characterize a given ghostless theory of massive gravity. The tensors $X_{\mu\nu}^{(1,2,3)}$ denote the interactions with the helicity-0 mode (de Rham and Gabadadze 2010b)

$$X_{\mu\nu}^{(1)} = \Box \pi g_{\mu\nu} - \Pi_{\mu\nu} \tag{1.54}$$

$$X_{\mu\nu}^{(2)} = \Pi_{\mu\nu}^2 - \Box \pi \Pi_{\mu\nu} - \frac{1}{2}([\Pi^2] - [\Pi]^2) g_{\mu\nu} \tag{1.55}$$

$$\begin{aligned} X_{\mu\nu}^{(3)} &= 6\Pi_{\mu\nu}^3 - 6[\Pi]\Pi_{\mu\nu}^2 + 3([\Pi]^2 - [\Pi^2])\Pi_{\mu\nu} \\ &\quad - g_{\mu\nu}([\Pi]^3 - 3[\Pi^2][\Pi] + 2[\Pi^3]) . \end{aligned} \tag{1.56}$$

Quite soon de Rham, Gabadadze and Tolley realized that beyond the decoupling limit these infinite interactions (1.47) could be resumed into a compact expression once the quantity

$$\mathcal{K}_\nu^\mu(g, H) = \delta_\nu^\mu - \sqrt{\delta_\nu^\mu - H_\nu^\mu} = -\sum_{n=1}^{\infty} d_n (H^n)_\nu^\mu \tag{1.57}$$

with

$$d_n = \frac{(2n)!}{(1-2n)(n!)^2 4^n} \tag{1.58}$$

and with the property

$$\mathcal{K}_\nu^\mu(g, H)|_{h_{\mu\nu}=0} = \Pi_{\mu\nu}, \tag{1.59}$$

[4]Note that we are neglecting the helicity-1 field in φ^a since at linear order this field decouples completely. If we had included the helicity-1 field in the expansion of $\varphi^a = (x^\alpha + A^\alpha - \pi^\alpha)\,\delta_\alpha^a$ then the expression for the covariant tensor would have contained the contribution of that field as well $H_{\mu\nu} = \frac{h_{\mu\nu}}{M_{Pl}} + \partial_\mu A_\nu + \partial_\nu A_\mu + \partial_\mu \pi_\nu + \partial_\nu \pi_\mu + \cdots$.

was defined (de Rham et al. 2011). They showed that the most generic potential that bears no Boulware-Deser ghost in four dimensions is

$$\mathcal{U}(g,\mathcal{H}) = -4\left(\mathcal{U}_2 + \alpha_3\,\mathcal{U}_3 + \alpha_4\,\mathcal{U}_4\right) \tag{1.60}$$

where $\alpha_{3,4}$ are two free parameters and the potentials U_i given by de Rham et al. (2011)

$$\mathcal{U}_2 = [\mathcal{K}]^2 - [\mathcal{K}^2] \tag{1.61}$$
$$\mathcal{U}_3 = [\mathcal{K}]^3 - 3[\mathcal{K}][\mathcal{K}^2] + 2[\mathcal{K}^3] \tag{1.62}$$
$$\mathcal{U}_4 = [\mathcal{K}]^4 - 6[\mathcal{K}^2][\mathcal{K}]^2 + 8[\mathcal{K}^3][\mathcal{K}] + 3[\mathcal{K}^2]^2 - 6[\mathcal{K}^4]\,. \tag{1.63}$$

These potential terms have been constructed by demanding the criteria that the $h = 0$ part of the interactions contain only total derivatives for the π-field. Soon it was realized that this criteria was automatically fulfilled by writing the interactions in terms of a deformed determinant (Hassan and Rosen 2011) $\det(\delta^\mu_\nu + \partial^\mu\partial_\nu\pi) = \sqrt{\det(g^{\mu\alpha}\mathrm{f}_{\alpha\nu})}|_{\mathrm{h}_{\mu\nu}=0}$. Since the determinant can be expressed in terms of the anti-symmetric Levi-Civita tensor

$$\det(\delta^\mu_\nu + \partial^\mu\partial_\nu\pi) = \sum_{i=0}^{4}\frac{-1}{\mathrm{i}!(4-\mathrm{i})!}\mathcal{E}_{\mu_1\cdots\mu_i\alpha_{i+1}\cdots\alpha_4}\mathcal{E}^{\nu_1\cdots\nu_i\alpha_{i+1}\cdots\alpha_4} \times \partial^{\mu_1}\partial_{\nu_1}\pi\cdots\partial^{\mu_i}\partial_{\nu_i}\pi \tag{1.64}$$

the total derivative nature of the interactions for the π-field is manifest. Beyond $\mathrm{h} = 0$ the mass term can be constructed by replacing $\delta^\mu_\nu + \partial^\mu\partial_\nu\pi$ by $\sqrt{g^{\mu\alpha}\mathrm{f}_{\alpha\nu}} = \delta^\mu_\nu + \mathrm{K}^\mu_\nu$ in the determinant

$$\det(\delta^\mu_\nu + \mathrm{K}^\mu_\nu) = \sum_{\mathrm{i}=0}^{4}\frac{-\alpha_\mathrm{i}}{\mathrm{i}!(4-\mathrm{i})!}\mathcal{E}_{\mu_1\cdots\mu_i\alpha_{i+1}\cdots\alpha_4}\mathcal{E}^{\nu_1\cdots\nu_i\alpha_{i+1}\cdots\alpha_4} \times \mathrm{K}^{\mu_1}_{\nu_1}\cdots\mathrm{K}^{\mu_i}_{\nu_i} \tag{1.65}$$

This can be written further in a more compact way realizing that the determinant of a matrix $\sqrt{g^{\mu\alpha}\mathrm{f}_{\alpha\nu}}$ can be expressed in terms of the elementary symmetric polynomials

$$\det(\delta^\mu_\nu + \mathrm{K}^\mu_\nu) = \sum_{\mathrm{n}=0}^{4}\alpha_\mathrm{n}\mathrm{e}_\mathrm{n}(\mathrm{K}) \tag{1.66}$$

with

$$
\begin{aligned}
e_0(K) &= 1\\
e_1(K) &= [K]\\
e_2(K) &= \frac{1}{2}([K]^2 - [K^2])\\
e_3(K) &= \frac{1}{6}([K]^3 - 3[K][K^2] + 2[K^3])\\
e_4(K) &= \frac{1}{24}([K]^4 - 6[K]^2[K^2] + 3[K^2]^2 + 8[K][K^3] - 6[K^4])
\end{aligned}
\tag{1.67}
$$

Notice that e_4 can be expressed in terms of $e_{2,3}$ and the tadpole $e_1 = [\mathcal{K}]$, Hassan and Rosen (2011). The potential interactions can be also in an analog way constructed from the polynomials

$$\sum_{n=0}^{4} \beta_n e_n(\sqrt{(g^{\mu\alpha} f_{\alpha\nu})}). \tag{1.68}$$

The Lagrangian for the massive graviton written in this compact form then becomes

$$\mathcal{L} = M_{Pl}^2 \sqrt{-g} R + 2M_{Pl}^2 m^2 \sqrt{-g} \sum_{n=0}^{4} \beta_n e_n(\sqrt{(g^{\mu\alpha} f_{\alpha\nu})}), \tag{1.69}$$

where the elementary symmetric polynomials can also be written in terms of the eigenvalues λ_i of $\sqrt{(g^{\mu\alpha} f_{\alpha\nu})}$ (Hassan and Rosen 2011)

$$
\begin{aligned}
e_0(\sqrt{(g^{\mu\alpha} f_{\alpha\nu})}) &= 1\\
e_1(\sqrt{(g^{\mu\alpha} f_{\alpha\nu})}) &= \lambda_1 + \lambda_2 + \lambda_3 + \lambda_4\\
e_2(\sqrt{(g^{\mu\alpha} f_{\alpha\nu})}) &= \lambda_1\lambda_2 + \lambda_1\lambda_3 + \lambda_1\lambda_4 + \lambda_2\lambda_3 + \lambda_2\lambda_4 + \lambda_3\lambda_4\\
e_3(\sqrt{(g^{\mu\alpha} f_{\alpha\nu})}) &= \lambda_1\lambda_2\lambda_3 + \lambda_1\lambda_2\lambda_4 + \lambda_1\lambda_3\lambda_4 + \lambda_2\lambda_3\lambda_4\\
e_4(\sqrt{(g^{\mu\alpha} f_{\alpha\nu})}) &= \lambda_1\lambda_2\lambda_3\lambda_4.
\end{aligned}
\tag{1.70}
$$

The forth polynomial

$$\sqrt{-g}\beta_4 e_4(\sqrt{(g^{\mu\alpha} f_{\alpha\nu})}) = \sqrt{-g}\beta_4 \det(\sqrt{(g^{-1} f)}) = \beta_4 \sqrt{\det f} \tag{1.71}$$

does not depend on $g_{\mu\nu}$ and therefore does not contribute to its equations of motion. In the case of a non-dynamical $f_{\mu\nu}$ this forth polynomial can be completely neglected. However, in the generalization of massive gravity to two dynamical metrics, meaning in the case in which both $g_{\mu\nu}$ and $f_{\mu\nu}$ are considered to be dynamical, the forth polynomial will correspond to a potential for $f_{\mu\nu}$ and can not any longer be neglected.

1.2.1 Bi-Gravity

This was precisely considered in Hassan and Rosen (2012) where they added an additional kinetic term for $f_{\mu\nu}$ and so constructed the first ghost-free non-linear bimetric theory for gravity

$$\mathcal{L} = M_{Pl}^2\sqrt{-g}R_g + M_f^2\sqrt{-f}R_f + 2M_{eff}^2 m^2\sqrt{-g}\sum_{n=0}^{4}\beta_n e_n(\sqrt{g^{-1}f}) \tag{1.72}$$

where

$$M_{eff}^2 = \left(\frac{1}{M_{Pl}^2} + \frac{1}{M_f^2}\right)^{-1}. \tag{1.73}$$

Note that the potential term is now the potential term for both metrics. The potential term can also be expressed in terms of the matrix $\sqrt{f^{-1}g}$ using the following relation

$$\sqrt{-\det g}\sum_{n=0}^{4}\beta_n e_n(\sqrt{g^{-1}f}) = \sqrt{-\det f}\sum_{n=0}^{4}\beta_n e_{4-n}(\sqrt{f^{-1}g}) \tag{1.74}$$

This bimetric theory at the linearized level corresponds to a massless spin-2 particle together with a massive spin-2 particle and hence contains seven physical degrees of freedom (Hassan and Rosen 2012). In order to make the mass spectrum apparent, consider the following metric perturbations

$$g_{\mu\nu} = \eta_{\mu\nu} + \frac{2}{M_{Pl}}h_{\mu\nu} \tag{1.75}$$

$$f_{\mu\nu} = \eta_{\mu\nu} + \frac{2}{M_f}l_{\mu\nu} \tag{1.76}$$

The action for the bigravity (1.72) reduces simply to

$$S = \int d^4x \left\{ -h_{\mu\nu}\hat{\mathcal{E}}^{\mu\nu\alpha\beta}h_{\alpha\beta} - l_{\mu\nu}\hat{\mathcal{E}}^{\mu\nu\alpha\beta}l_{\alpha\beta} \right.$$
$$\left. -\frac{m^2M_{eff}^2}{4}\left[\left(\frac{h^\mu_\nu}{M_{Pl}} - \frac{l^\mu_\nu}{M_f}\right)^2 - \left(\frac{h^\mu_\mu}{M_{Pl}} - \frac{l^\mu_\mu}{M_f}\right)^2\right]\right\}, \tag{1.77}$$

with M_{eff}^2 as defined in (1.73). We can diagonalize these interactions by making the following change of variables

$$\frac{1}{M_{eff}} w_{\mu\nu} = \frac{h_{\mu\nu}}{M_f} + \frac{l_{\mu\nu}}{M_{Pl}}$$
$$\frac{1}{M_{eff}} v_{\mu\nu} = \frac{h_{\mu\nu}}{M_{Pl}} - \frac{l_{\mu\nu}}{M_f}, \tag{1.78}$$

such that the action at linear order becomes (Hassan and Rosen 2012)

$$S = \int d^4x \left\{ -w_{\mu\nu}\hat{\mathcal{E}}^{\mu\nu\alpha\beta} w_{\alpha\beta} - v_{\mu\nu}\hat{\mathcal{E}}^{\mu\nu\alpha\beta} v_{\alpha\beta} - \frac{m^2}{4}\left[[v^2] - [v]^2 \right] \right\}. \tag{1.79}$$

In the unitary gauge, $v_{\mu\nu}$ encodes all the five physical degrees of freedom of a massive spin-2 fluctuation, namely the two helicity-2, the two helicity-1 and the helicity-0 mode, and $w_{\mu\nu}$ encodes the two helicity-2 modes of the massless spin-2 fluctuation.

1.2.2 Vielbein Formulation

In the metric formulation the potential term for the massive gravity (as well as for the bimetric gravity) has the complication through the matrix square roots. This mathematically cumbersome structure can be avoided using the vielbein $\mathcal{E}^A = \mathcal{E}_\mu{}^A$ instead of the metric since the vielbein is the "square root" of the metric $g_{\mu\nu} = \mathcal{E}_\mu{}^A E_\nu{}^B \eta_{AB}$. In the vielbein language, the Lagrangian for the massive gravity becomes

$$\begin{aligned} \mathcal{L} = M_{Pl}^2 \mathcal{E}_{abcd} \Bigg[&\frac{1}{4}\mathcal{E}^a \wedge \mathcal{E}^b \wedge R^{cd} - \frac{m^2\beta_0}{4!}\mathcal{E}^a \wedge \mathcal{E}^b \wedge \mathcal{E}^c \wedge \mathcal{E}^d \\ &- \frac{m^2\beta_1}{3!} I^a \wedge \mathcal{E}^b \wedge \mathcal{E}^c \wedge \mathcal{E}^d - \frac{m^2\beta_2}{2!2!} I^a \wedge I^b \wedge \mathcal{E}^c \wedge \mathcal{E}^d \\ &- \frac{m^2\beta_3}{3!} I^a \wedge I^b \wedge I^c \wedge \mathcal{E}^d - \frac{m^2\beta_4}{4!} I^a \wedge I^b \wedge I^c \wedge I^d \Bigg] \end{aligned} \tag{1.80}$$

where $I^a = \delta^a_\mu dx^\mu$ (Hinterbichler and Rosen 2012; Ondo and Tolley 2013). In a very similar way as in the metric formulation one can restore gauge invariance by adding the Stueckelberg fields

$$\mathcal{E}^a_\mu \to \Lambda^a{}_b \mathcal{E}^b_\nu \frac{\partial x^\nu}{\partial \phi^\mu} \tag{1.81}$$

Along the lines of the vielbein formulation the bimetric gravity similarly becomes

$$\begin{aligned}\mathcal{L} = M_{Pl}^2 \mathcal{E}_{abcd} \Bigg[& \frac{1}{4}\mathcal{E}^a \wedge \mathcal{E}^b \wedge R_{\mathcal{E}}^{cd} + \frac{1}{4}F^a \wedge F^b \wedge R_F^{cd} \\ & - \frac{m^2\beta_0}{4!}\mathcal{E}^a \wedge \mathcal{E}^b \wedge \mathcal{E}^c \wedge \mathcal{E}^d - \frac{m^2\beta_1}{3!}F^a \wedge \mathcal{E}^b \wedge \mathcal{E}^c \wedge \mathcal{E}^d \\ & - \frac{m^2\beta_2}{2!2!}F^a \wedge F^b \wedge \mathcal{E}^c \wedge \mathcal{E}^d - \frac{m^2\beta_3}{3!}F^a \wedge F^b \wedge F^c \wedge \mathcal{E}^d \\ & - \frac{m^2\beta_4}{4!}F^a \wedge F^b \wedge F^c \wedge F^d \Bigg]\end{aligned} \tag{1.82}$$

where F^a is the vielbein corresponding to the metric f as $f_{\mu\nu} = F_\mu{}^A F_\nu{}^B \eta_{AB}$ (Hinterbichler and Rosen 2012; Fasiello and Tolley 2013). The equivalence between the vielbein and metric formulation can be shown using the symmetric vielbein condition $(\Lambda E)^a_\mu \eta_{ab} F^b_\nu = (\Lambda \mathcal{E})^b_\mu \eta_{ab} F^a_\nu$ (Hinterbichler and Rosen 2012; Ondo and Tolley 2013; Hassan et al. 2012).

The ghost free construction of massive gravity has initiated a flurry of investigations in the field. Not only its extension to bi- and multi metric gravity theories but also the question of whether or not it might be at the origin of the observed accelerated expansion of the universe has woken up a lot of interests. It is also going to be the central topic of this thesis. We will try to address the burden problems of modern cosmology like the cosmological constant problem or the dark energy problem in the framework of massive gravity and discuss the reliability of its solutions in given specific realizations.

1.3 Galileons

Another important class of infra-red modifications is the Galileon theory. As we mentioned before the Galileon interactions were introduced as a natural extension of the decoupling limit of DGP (Nicolis et al. 2009). It can be considered as an effective field theory constructed by the restriction of the invariance under internal Galilean and shift transformations and the ghost absence. In order for the theory to be viable, the Vainshtein mechanism is needed, which on the other hand relies on the presence of interactions at an energy scale $\Lambda \ll M_{Pl}$. From a traditional effective field theory point of view these interactions are irrelevant operators which renders the theory non-renormalizable, but to contrary to the traditional case, within the Galileon theory these irrelevant operators need to be large in the regime of interest, in the so called strong coupled regime $\partial^2\pi \sim \Lambda^3$. Therefore, one might have concerns that the effective field theory could go out of control in this strong coupling regime where the irrelevant operators need to be large. Nevertheless, the Galileon theories are not typical effective field theories in the sense that it is organized in the small parameter expansion of the whole operator but rather it has to be reorganized in a way that the derivative now plays the role of the small parameter rather than the whole operator itself. There exist a regime of interest for which $\pi \sim \Lambda$, $\partial\pi \sim \Lambda^2$ and $\partial^2\pi \sim \Lambda^3$ even though any further derivative is suppressed $\partial^3\pi \ll \Lambda^4$, meaning that the effective field expansion is reorganized such that the Galileon interactions are the relevant operators with equations of motion with only two derivatives, while all other interactions with equations of motion with more than two derivatives are treated as negligibale corrections.

The symmetry and ghost absence conditions are fulfilled by only five interactions in four dimensions:

$$\begin{aligned}
\mathcal{L}_1 &= \pi \\
\mathcal{L}_2 &= (\partial\pi)^2 \\
\mathcal{L}_3 &= (\partial\pi)^2\Box\pi \\
\mathcal{L}_4 &= (\partial\pi)^2\left[(\Box\pi)^2 - (\partial_\mu\partial_\nu\pi)^2\right] \\
\mathcal{L}_5 &= (\partial\pi)^2\left[(\Box\pi)^3 - 3\Box\pi(\partial_\mu\partial_\nu\pi)^2 + 2(\partial_\mu\partial_\nu\pi)^3\right]
\end{aligned} \tag{1.83}$$

This effective action is local and contains higher order derivatives. Nevertheless, these interactions come in a very specific way such that they only give rise to second order equations of motion, which are invariant under the transformations

$$\pi \to \pi + c + x_\mu b^\mu \tag{1.84}$$

with constant c and b^μ. They can be written in a more compact way by using the iterative relation

$$\mathcal{L}_{n+1} = -(\partial\pi)^2\mathcal{E}_n \tag{1.85}$$

with $n \geq 1$ and $\mathcal{E}_n = \frac{\delta\mathcal{L}_n}{\delta\pi}$ are the equations of motion

$$\begin{aligned}
\mathcal{E}_1 &= 1 \\
\mathcal{E}_2 &= \Box\pi \\
\mathcal{E}_3 &= (\Box\pi)^2 - (\partial_\mu\partial_\nu\pi)^2 \\
\mathcal{E}_4 &= (\Box\pi)^3 - 3\Box\pi(\partial_\mu\partial_\nu\pi)^2 + 2(\partial_\mu\partial_\nu\pi)^3 \\
\mathcal{E}_5 &= (\Box\pi)^4 - 6(\Box\pi)^2(\partial_\mu\partial_\nu\pi)^2 + 8\Box\pi(\partial_\mu\partial_\nu\pi)^3 + 3((\partial_\mu\partial_\nu\pi)^2)^2 \\
&\quad -6(\partial_\mu\partial_\nu\pi)^4
\end{aligned} \tag{1.86}$$

The allowed interactions for the Galileon were originally determined order by order by writing down all the possible contractions for the scalar field interactions and without giving any higher dimensional realization, for which these theories could be the decoupling limit of. However, de Rham and Tolley could construct an unified class of four dimensional effective theories starting from a higher dimensional setup and show that these effective theories reproduce successfully all the interaction terms of the Galileon in the non-relativistic limit (de Rham and Tolley 2010). The construction of the action is based on a brane localized in a higher dimensional bulk $y = \pi(x_\mu)$ where x_μ are the four-dimensional coordinates while y describes the direction of the fifth dimension. The planck masses of the four and five dimensions are M_4 and M_5 respectively and their ratio is $m = M_5^3/M_4^2$. The Lagrangian is allowed to contain interactions which only yield second order equations of motion and which are manifestly covariant. The Lovelock invariants are the only manifestly covariant terms giving rise to second order equations of motion, therefore the allowed interactions are the Lovelock invariants in four dimensions or the Gibbons-Hawking-York boundary terms associated with Lovelock invariants in five dimensions (de Rham and Tolley 2010).

$$\mathcal{L} = \sqrt{-g}\left(-\lambda + \frac{M_4^2}{2}R - M_5^3 K - \beta\frac{M_5^3}{m^2}\mathcal{K}_{GB}\right) \tag{1.87}$$

where $K_{\mu\nu}$ is the extrinsic curvature on the brane (K its trace) and $\mathcal{K}_{GB}$ is the Gibbons-Hawking-York boundary term

$$\mathcal{K}_{GB} = -\frac{2}{3}K^3_{\mu\nu} + KK^2_{\mu\nu} - \frac{1}{3}K^3 - 2G_{\mu\nu}K^{\mu\nu} \tag{1.88}$$

associated with a bulk Gauss-Bonnet term (Davis 2003; de Rham and Tolley 2010)

$$\mathcal{L}_{GB} = R^2 + 2R^2_{\mu\nu\alpha\beta} - 4R^2_{\mu\nu}. \tag{1.89}$$

In four dimensional space-time the only non-trivial Lovelock invariants are the cosmological constant and the Ricci scalar. Related to the Lovelock invariants one can construct divergenceless tensors. The divergeless tensor associated to the cosmological constant is the metric $g_{\mu\nu}$ and the divergence less tensor associated to the ricci scalar is the einstein tensor $G_{\mu\nu}$. Since the Gauss Bonnet Lovelock invariant in four dimensions is just a total derivative the corresponding divergeceless tensor is zero. However, one can still construct another divergeceless tensor, which is the dual Riemann tensor $\mathcal{L}_{\mu\alpha\nu\beta}$ and linear in the curvature. Written in terms of the Levi-Civita antisymmetric tensors their structure is more apparent (linearized in perturbation $h_{\mu\nu}$) (Fig. 1.5).

$$\begin{aligned}
\lambda &\sim \mathcal{E}^{abcd}\mathcal{E}_{abcd} \\
g_{\mu\nu} &\sim \mathcal{E}_{\mu}{}^{abc}\mathcal{E}_{\nu abc} \\
R &\sim \mathcal{E}^{abcd}\mathcal{E}^{a'b'}{}_{cd}\,\partial_a\partial_{a'}h_{bb'} \\
G_{\mu\nu} &\sim \mathcal{E}_{\mu}{}^{abc}\mathcal{E}_{\nu}{}^{a'b'}{}_{c}\,\partial_a\partial_{a'}h_{bb'} \\
L_{\mu\alpha\nu\beta} &\sim \mathcal{E}_{\mu\alpha}{}^{ab}\mathcal{E}_{\nu\beta}{}^{a'b'}\,\partial_a\partial_{a'}h_{bb'}
\end{aligned} \tag{1.90}$$

However, in five dimensions we can have an additional divergenceless tensor like the dual Riemann tensor but this time with six indices (still linear in the curvature).

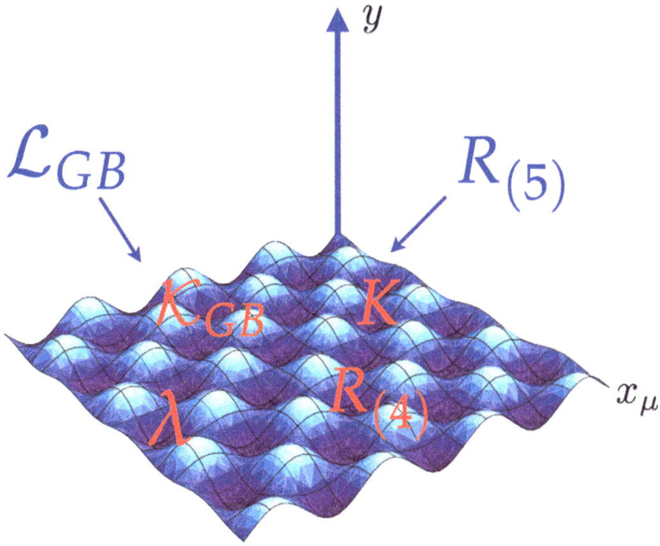

Fig. 1.5 A four dimensional brane embedded in five dimensional bulk: the Lovelock invariants on the brane are the cosmological constant λ and four dimensional Ricci scalar $R(4)$. The Lovelock invariants in five dimensions are the five dimensional Ricci scalar $R(5)$ and the Gauss Bonnet term $\mathcal{L}_{GB}$, which give rise to the extrinsic curvature K and Gibbons-Hawking-York boundary term on the brane K_{GB}

$$
\begin{aligned}
\lambda &\sim \mathcal{E}^{abcde}\mathcal{E}_{abcde} \\
g_{\mu\nu} &\sim \mathcal{E}_{\mu}{}^{abcd}\mathcal{E}_{\nu abcd} \\
R &\sim \mathcal{E}^{abcde}\mathcal{E}^{a'b'}{}_{cde}\partial_a\partial_{a'}h_{bb'} \\
G_{\mu\nu} &\sim \mathcal{E}_{\mu}{}^{abcd}\mathcal{E}_{\nu}{}^{a'b'}{}_{cd}\partial_a\partial_{a'}h_{bb'} \\
\mathcal{L}_{\mu\alpha\nu\beta} &\sim \mathcal{E}_{\mu\nu}{}^{abc}\mathcal{E}_{\alpha\beta}{}^{a'b'}{}_{c}\partial_a\partial_{a'}h_{bb'} \\
L_{\mu\alpha\nu\beta\rho\sigma} &\sim \mathcal{E}_{\mu\nu\rho}{}^{ab}\mathcal{E}_{\alpha\beta\sigma}{}^{a'b'}\partial_a\partial_{a'}h_{bb'}
\end{aligned}
\tag{1.91}
$$

The Gauss Bonnet lovelock invariant, being second order in the curvature, corresponds to a total derivative in four dimensions and therefore does not yield any non-trivial contribution. However, in five dimensions it yields a non-trivial contribution and therefore there exists an additional divergenceless tensor associated to the Gauss Bonnet lovelock invariant, which can be expressed as

$$
\mathcal{L}^{GB}_{\mu\nu} = \mathcal{R}_{\mu\nu} - \frac{1}{4}\mathcal{R}g_{\mu\nu}, \tag{1.92}
$$

where $\mathcal{R}_{\mu\nu}$ stands for the short-cut notation

$$\mathcal{R}_{\mu\nu} = RR_{\mu\nu} - 2R_{\mu\alpha}R^{\alpha}_{\nu} - 2R^{\alpha\beta}R_{\mu\alpha\nu\beta} + R_{\mu\alpha\beta\rho}R^{\alpha\beta\rho}_{\nu}. \tag{1.93}$$

Again written in terms of the Levi-Civita antisymmetric tensor for a small fluctuation $h_{\mu\nu}$ the structure becomes more apparent

$$\begin{aligned} \mathcal{L}_{GB} &= \mathcal{E}^{abcde}\mathcal{E}^{a'b'c'd'}{}_{e}\partial_a\partial_{a'}h_{bb'}\partial_c\partial_{c'}h_{dd'} \\ \mathcal{L}^{GB}_{\mu\nu} &= \mathcal{E}_{\mu}{}^{abcde}\mathcal{E}_{\nu}{}^{a'b'c'd'}\partial_a\partial_{a'}h_{bb'}\partial_c\partial_{c'}h_{dd'}. \end{aligned} \tag{1.94}$$

In the following, we will consider a probe brane embedded in five dimensional Minkowski space-time at the position $y = \pi(x_\mu)$. First of all, the induced metric on the brane will be given by

$$\begin{aligned} g_{\mu\nu} &= \eta_{\mu\nu} + \partial_\mu\pi\partial_\nu\pi \\ g^{\mu\nu} &= \eta^{\mu\nu} - \frac{\partial^\mu\pi\partial^\nu\pi}{1+(\partial\pi)^2} \end{aligned} \tag{1.95}$$

while the extrinsic curvature by

$$K_{\mu\nu} = -\gamma\partial_\mu\partial_\nu\pi \qquad \text{with} \qquad \gamma = \frac{1}{\sqrt{1+(\partial\pi)^2}}. \tag{1.96}$$

The individual invariants in (1.87) on the brane are the cosmological constant $\mathcal{L}^f_2$, the extrinsic curvature $\mathcal{L}^f_3$, the induced Ricci tensor $\mathcal{L}^f_4$ and the boundary term from the Gauss-Bonnet curvature in the bulk $\mathcal{L}^f_5$. Thus, the invariants (1.87) on the flat background become

$$\begin{aligned} \mathcal{L}^f_2 &= \gamma^- 1 \\ \mathcal{L}^f_3 &= \left([\Pi] - \gamma^2[\phi]\right) \\ \mathcal{L}^f_4 &= \gamma\left(([\Pi]^2 - [\Pi^2]) + 2\gamma^2([\phi^2] - [\Pi][\phi])\right) \\ \mathcal{L}^f_5 &= \gamma^2\Big(([\Pi]^3 + 2[\Pi^3] - 3[\Pi][\Pi^2]) + 6\gamma^2([\Pi][\phi^2] - [\phi^3]) \\ &\quad - 3\gamma^2([\Pi]^2 - [\Pi^2])[\phi]\Big), \end{aligned} \tag{1.97}$$

where $\Pi_{\mu\nu} = \partial_\mu\partial_\nu\pi$ and $[\phi^n] = \partial\pi \cdot \Pi^n \cdot \partial\pi$. Taking now the non-relativistic limit $(\partial\pi)^2 \ll 1$ gives exactly the Galileon interactions in (1.83) back[5] (de Rham and Tolley 2010). Galileon theories have also initiated a lava of investigations in cosmology. Not only the fact that they give rise to second order equations of motion

[5]In general, there is also a tadpole contribution $\mathcal{L}_1$, that depends on the bulk content.

and have symmetry, but also the non-renormalization theorem makes them theoretically very interesting since once the parameters in the theory are tuned by observational constraints they are going to be stable under quantum corrections. Even if the naive covariantization of the Galileon interactions on non-flat backgrounds break the Galileon symmetry explicitly, one can successfully generalize the Galileon interactions to the case with maximally symmetric backgrounds. This was precisely what we studied in our work in Burrage et al. (2011). For this, consider now a probe brane embedded into a warped five dimensional de Sitter space-time instead of Minkowski as before

$$ds_5^2 = e^{-2Hy}\left(dy^2 + q_{\mu\nu}dx^\mu dx^\nu\right), \tag{1.98}$$

where $q_{\mu\nu}$ is no longer flat as in Eq. (1.95) but rather taken to be in a de Sitter slicing

$$ds_{dS}^2 = q_{\mu\nu}dx^\mu dx^\nu = -n^2(t)dt^2 + a^2(t)dx^2 \tag{1.99}$$

with $\frac{\dot{a}}{na} = \text{const} = H$. The brane is still positioned at $y = \pi(x_\mu)$. Now again, from the five-dimensional theory, five invariant quantities can be induced on the brane (de Rham and Tolley 2010): The tadpole $\mathcal{L}_1^{dS}$, the DBI equivalent $\mathcal{L}_2^{dS}$, the extrinsic curvature $\mathcal{L}_3^{dS}$, the induced Ricci tensor $\mathcal{L}_4^{dS}$ and finally the boundary term from the Gauss-Bonnet curvature in the bulk $\mathcal{L}_5^{dS}$. These are constructed out of the induced metric at $y = \pi(x^\mu)$

$$g_{\mu\nu} = e^{-2H\pi}\left(q_{\mu\nu} + \partial_\mu\pi\partial_\nu\pi\right). \tag{1.100}$$

For a de Sitter geometry the invariants are

$$\begin{aligned}
\sqrt{-q}\mathcal{L}_1^{dS} &= \frac{1}{5H}\sqrt{-q}\left(e^{-5H\pi} - 1\right)\\
\sqrt{-q}\mathcal{L}_2^{dS} &= \sqrt{-g} = e^{-4H\pi}\sqrt{-q}\mathcal{L}_2^{f}\\
\sqrt{-q}\mathcal{L}_3^{dS} &= -\sqrt{-g}K = e^{-3H\pi}\sqrt{-q}\left(\mathcal{L}_3^{f} + 4H\gamma\mathcal{L}_2^{f}\right)\\
\sqrt{-q}\mathcal{L}_4^{dS} &= \sqrt{-g}R = e^{-2H\pi}\sqrt{-q}\left(\mathcal{L}_4^{f} + 6H\gamma\mathcal{L}_3^{f} + 12H^2\gamma^2\mathcal{L}_2^{f}\right)\\
\sqrt{-q}\mathcal{L}_5^{dS} &= -\frac{3}{2}\sqrt{-g}\,\mathcal{K}_{GB}\\
&= e^{-H\pi}\sqrt{-q}\left(\mathcal{L}_5^{f} + 6H\gamma\mathcal{L}_4^{f} + 18H^2\gamma^2\mathcal{L}_3^{f}\right.\\
&\quad \left. + 24H^3\gamma^3\mathcal{L}_2^{f}\right)
\end{aligned} \tag{1.101}$$

where the $\mathcal{L}_i^{f}$ are the flat space invariants (1.97) and K_{GB} is the Gibbons-Hawking-York boundary term as in Eq. (1.88) associated with a bulk Gauss-Bonnet term (1.89), with the expressions for the extrinsic and intrinsic curvature on the brane are given by the following,

$$K_{\mu\nu} = \gamma e^{-H\pi}\left(\Pi_{\mu\nu} + H\gamma_{\mu\nu} + H\partial_\mu\pi\partial_\nu\pi\right) \tag{1.102}$$

$$\begin{aligned} R_{\mu\nu} &= 2H\gamma^2\Pi_{\mu\nu} + \gamma^2([\Pi]\Pi_{\mu\nu} - \Pi^2_{\mu\nu}) + \gamma^4(\phi^2_{\mu\nu} - [\phi]\Pi_{\mu\nu}) \\ &\quad + H\gamma^2\left(3H + \gamma^2[\Pi] + \gamma^2([\Pi](\partial\pi)^2 - [\phi])\right)(q_{\mu\nu} + \partial_\mu\pi\partial_\nu\pi) \end{aligned} \tag{1.103}$$

$$\begin{aligned} R &= e^{2H\pi}\Big(12H^2\gamma^2 + 6H\gamma^2\left([\Pi] - \gamma^2[\phi]\right) + \gamma^2([\Pi]^2 - [\Pi^2]) \\ &\quad + 2\gamma^4([\phi^2] - [\phi][\pi])\Big). \end{aligned} \tag{1.104}$$

Furthermore they also satisfy the recursive relations, first introduced by Fairlie et al. (1992), and explained more recently within the context of the Galileon (de Rham and Tolley 2010; Fairlie 2011)

$$\mathcal{L}^{dS}_{n+1} = -e^{H\pi}\gamma^{-1}\frac{\delta\mathcal{L}^{dS}_n}{\delta\pi} \qquad \text{for} \quad n \geq 1\,, \tag{1.105}$$

which generalizes the flat space-time relations (1.85). It is straightforward to check that there cannot be any further invariant beyond $n = 5$ because $\delta_\pi\mathcal{L}^{dS}_5$ is a total derivative. Similarly as in the flat space case, we can build the non-relativistic limit, $(\partial\pi)^2 \ll 1$, of this theory. For this, we first canonically normalize the field $\pi = \hat{\pi}/\sqrt{\lambda}$ and then send $\lambda \to \infty$. In this limit the Galileon symmetry on flat background (1.84) is promoted to (after setting the lapse to $n = 1$),

$$\hat{\pi} \longrightarrow \hat{\pi} + e^{Ht}\left(c + v_i x^i + \frac{1}{2}v_0 H x^i x_i\right) + \frac{v_0}{H}\sinh(Ht)\,, \tag{1.106}$$

which in the flat space limit $H \to 0$ reduces to the Galileon shift symmetry $\hat{\pi} \to \hat{\pi} + \left(c + v_\mu x^\mu\right)$ of Nicolis et al. (2009). Keeping the lapse arbitrary to retain the gauge freedom, the Galilean transformation becomes,

$$\hat{\pi} \longrightarrow \hat{\pi} + a(t)\left(c + v_i x^i + v_0 H x^i x_i - v_0\left(\frac{n(t)}{\dot{a}(t)}\right)^2\right), \tag{1.107}$$

with $\frac{\dot{a}}{an} = H = \text{const}$.

The specific combination of $\mathcal{L}^{dS}_1$ and $\mathcal{L}^{dS}_2$ which remains finite in the limit $\lambda \to \infty$ is

$$\mathcal{L}^{(NR)}_2 = \lambda(\mathcal{L}^{dS}_2 + 4\mathcal{L}^{dS}_1 - 1) \xrightarrow{\lambda\to\infty} \tfrac{1}{2}\left((\partial\hat{\pi})^2 - 4H^2\hat{\pi}^2\right), \tag{1.108}$$

which is indeed invariant under the non-relativistic transformation (1.106), and which gives back the usual first Galileon kinetic term, $1/2(\partial\pi)^2$, in the limit $H \to 0$.

We can now go further and consider the non-relativistic limit of the higher order invariants (extrinsic curvature term, scalar curvature, etc.) which are also invariant under (1.106). The de Sitter generalization of the Galileon derivative interactions are then:

$$\begin{aligned}
\mathcal{L}_2^{(NR)} &= (\partial\hat{\pi})^2 - 4H^2\hat{\pi}^2 \\
\mathcal{L}_3^{(NR)} &= (\partial\hat{\pi})^2\Box\hat{\pi} + 6H^2\hat{\pi}(\partial\hat{\pi})^2 - 8H^4\hat{\pi}^3 \\
\mathcal{L}_4^{(NR)} &= (\partial\hat{\pi})^2\left([\hat{\Pi}^2] - [\hat{\Pi}]^2\right) - 6H^2\hat{\pi}(\partial\hat{\pi})^2\Box\hat{\pi} \\
&\quad -\frac{1}{2}H^2(\partial\hat{\pi})^4 - 18H^4\hat{\pi}^2(\partial\hat{\pi})^2 + 12H^6\hat{\pi}^4 \\
\mathcal{L}_5^{(NR)} &= (\partial\hat{\pi})^2\left([\hat{\Pi}]^3 - 3[\hat{\Pi}^2][\hat{\Pi}] + 2[\hat{\Pi}^3]\right) - 4H^2\hat{\pi}(\partial\hat{\pi})^2\left([\hat{\Pi}^2] - [\hat{\Pi}]^2\right) \\
&\quad -\frac{1}{2}H^2(\partial\hat{\pi})^4\Box\hat{\pi}12H^4\hat{\pi}^2(\partial\hat{\pi})^2\Box\hat{\pi} + 2H^4\pi(\partial\hat{\pi})^4 \\
&\quad + 24H^6\hat{\pi}^3(\partial\hat{\pi})^2 - \frac{48}{5}H^8\hat{\pi}^5 . \qquad (1.109)
\end{aligned}$$

These are the de Sitter Galileons. One can check that the actions $S_i^{(NR)} = \int a^3(t)n(t)\mathcal{L}_i^{(NR)}$ are indeed invariant under the transformation (1.107) up to a total derivative.

Similarly, one can consider a brane embedded into a warped five dimensional Anti de Sitter space-time. In this case, the induced metric and the extrinsic curvature on the brane become

$$\begin{aligned}
g_{\mu\nu} &= e^{-2\pi/l}\eta_{\mu\nu} + \partial_\mu\pi\partial_\nu\pi \\
K_{\mu\nu} &= -\tilde{\gamma}\left(\partial_\mu\partial_\nu\pi + \frac{\partial_\mu\pi\partial_\nu\pi}{l} + \frac{g_{\mu\nu}}{l}\right) \qquad (1.110)
\end{aligned}$$

where l is AdS length and $\tilde{\gamma} = \frac{1}{\sqrt{1+e^{2\pi/l}(\partial\pi)^2}}$. Now, taking the non-relativistic limit $e^{2\pi/l}(\partial\pi)^2 \ll 1$ yields the conformal extension of the Galilean. This is equivalent to defining $\pi = l\hat{\pi}$ and then taking the limit $l \to 0$. The conformal Galilean have the following form (de Rham and Tolley 2010)

$$\begin{aligned}
\mathcal{L}_1 &= e^{-4\hat{\pi}} \\
\mathcal{L}_2 &= -\frac{1}{2}e^{-2\hat{\pi}}(\partial\hat{\pi})^2 \\
\mathcal{L}_3 &= \frac{1}{2}(\partial\hat{\pi})^2\Box\hat{\pi} - \frac{1}{4}(\partial\hat{\pi})^4 \\
\mathcal{L}_4 &= \frac{1}{20}e^{2\hat{\pi}}(\partial\hat{\pi})^2\left(10([\hat{\Pi}]^2 - [\hat{\Pi}^2]) + 4((\partial\hat{\pi})^2\Box\hat{\pi} - [\hat{\phi}]) + 3(\partial\hat{\pi})^4\right) \\
\mathcal{L}_5 &= e^{4\hat{\pi}}(\partial\hat{\pi})^2\left(\frac{1}{3}([\hat{\Pi}]^3 + 2[\hat{\Pi}^3] - 3[\hat{\Pi}][\hat{\Pi}^2]) + (\partial\hat{\pi})^2([\hat{\Pi}]^2 - [\hat{\Pi}^2])\right. \\
&\quad \left. + \frac{10}{7}(\partial\hat{\pi})^2((\partial\hat{\pi})^2[\hat{\Pi}] - [\hat{\phi}]) + \frac{1}{28}(\partial\hat{\pi})^6\right) . \qquad (1.111)
\end{aligned}$$

Recall that we are using the notation $(\partial\pi)^2 = \partial_\mu\pi\partial^\mu\pi$ and $\Pi_{\mu\nu} = \partial_\mu\partial_\nu\pi$ and $[\phi] = \partial_\mu\pi\partial_\nu\pi\partial^\mu\partial^\nu\pi$. Since Galileons and conformal Galileons have been constructed from a higher dimensional framework in which the Lovelock interactions manifestly fulfill the symmetry and give rise to only second order equations, the Galileon/conformal Galileons themselves give only second order equations of motion and yet preserve the symmetry. The generalization of the Galileon interactions we have inferred is only preserved on a pure de Sitter/Anti de Sitter background. However, for any cosmological scenario, we would need to include small departures from de Sitter. One could in principle covariantize this model, similarly as what was performed in Deffayet et al. (2009), however we would then loose the Galileon symmetry. Instead, another way forward is to generalize the description to a generic Friedmann-Robertson-Walker (FRW) background. In principle, this extension is straightforward when considering the FRW-slicing of five-dimensional Minkowski

$$ds^2 = e^{-2Hy}\Big[a^2(t)dx^2 + dy^2 + 2(1 + Hy - e^{Hy})\frac{\dot{H}}{nH^2}dydt \\ -(1 - \frac{\dot{H}}{nH})\left(1 + 2(1 - e^{Hy})\frac{\dot{H}}{nH^2} + y\frac{\dot{H}}{nH}\right)dt^2\Big]. \quad (1.112)$$

However, the realization of the Poincaré symmetry in this gauge is highly non-trivial, and its non-relativistic limit appears non-local. This is however not surprising, as there is no reasons why FRW which is not maximally symmetric should enjoy similar amount of symmetry. In the five-dimensional picture, the coordinate transformation to transfer from flat slicing of Minkowski to an FRW slicing is non-local, and so the Poincaré symmetry expressed in the five-dimensional FRW slicing does not have the similar close form as in de Sitter.

In this thesis we will not aim to study the Galileon interactions on FRW backgrounds and their cosmological impact but rather concentrate on their behavior around static spherically symmetric backgrounds. In particular, we will study the propagation speed of fluctuations on top of these backgrounds. We will devote some time to investigate how the superluminal propagation is an unavoidable feature in these Galileon theories.

1.4 Outline of This Work

This thesis presents the summary of the scientific results and knowledge gathered during my four years of PhD education. It consists of four parts and each part is further divided into chapters. In the first chapter we introduced the framework within which we will be working in this thesis, so at this stage we are in possession of all the important quantities for the remaining parts of the thesis. The first chapter was dedicated to the concept of field theories in cosmology, where we introduced the main features of the Standard Model of Particle Physics and the Standard Model of cosmology, paying special attention to the cosmic acceleration problem which became firmly established by means of a variety of cosmological observations. Moreover, we have summarized the three categories of the theoretical proposals for dark energy that have been considered in the literature and very soon concentrated on massive gravity as an alternative to dark energy in Sect. 1.2. We discussed the recently developed ghost-free nonlinear theory for massive spin-2 fields, the de Rham-Gabadadze-Tolley theory. We introduced the bimetric gravity model and illustrated its construction from massive gravity. Section 1.3 was devoted to the introduction of the Galileon interactions as an important class of infrared modifications of general relativity. We presented their main features and discussed how they can be constructed in the framework of higher dimensional space-time.

The first part of the thesis concentrates on the cosmology in the framework of massive gravity and consists of two chapters. Chapter 2 is the summary of our work in de Rham et al. (2011) where we study at great length the cosmology of the dRGT theory in the decoupling limit. In this chapter we will explore the existence of self-accelerating and degravitating solutions in the decoupling limit of massive gravity. We will put constraints on the parameters of the theory in the decoupling limit based on the classical stability in the cosmological evolution. From the decoupling limit we will construct a proxy theory to massive gravity in Chap. 3, which will represent our work in de Rham and Heisenberg (2011). This proxy theory corresponds to a very specific type of non-minimally coupled scalar-tensor interactions as a subclass of Horndeski theories. We will study the self-accelerating and degravitating solutions in this proxy theory as well. Furthermore, we will mention the analog non-minimal interactions for a vector field based on our work in Jiménez et al. (2013). This will give us also the opportunity to present our preliminary results on the Horndeski Proca field interactions, which describe the most general interactions for a vector field with three propagating degrees of freedom. We will finalize Chap. 3 with a summary and a critical view of our previous analysis and the assumptions made there.

The second part consisting of Chap. 4 discusses the superluminal propagation in Galileon models. This superluminal propagation is a shared property also in massive gravity since the Galileon models naturally arise in the decoupling limit of massive gravity. This chapter is manly based on our work in de Fromont et al. (2013) where we show in great detail that the feature of superluminal propagating solutions for multi-galileon theories is unavoidable.

The third part of the thesis is fully consecrated to the study of quantum corrections in massive gravity. It is a mandatory question to ask whether the parameters of the theory are stable under quantum corrections. We will start with the non-renormalization theorem in the decoupling limit and show how it protects the graviton mass from quantum corrections in Chap. 5. This will be the summary of our work in de Rham et al. (2012). We will then move on to explore the quantum corrections beyond the decoupling limit in Chap. 6 and study explicitly the stability of the graviton potential when including matter and graviton loops based on our work in de Rham et al. (2015). The analysis of the one-loop matter quantum corrections reveals that the potential remains unaffected since they contribute only in form of a cosmological constant. On the other hand, the one-loop quantum corrections coming from the gravitons do destabilize the special structure of the potential, howbeit even in the case of large background configuration, the Vainshtein mechanism redresses the one-loop effective action so that the detuning remains irrelevant below the Planck scale. This allows us to draw the conclusion that the one-loop quantum corrections to the potential are harmless. We will finish the Chap. 6 by presenting our preliminary results on the quantum corrections in bimetric gravity theories and commenting on some prospects concerning future potential investigations.

The last part of the thesis with Chap. 7 recapitulates the main results and contributions made in this thesis. We will also provide an outlook for future investigations in the field.

References

Arkani-Hamed N, Dimopoulos S, Dvali G, Gabadadze G (2002) Nonlocal modification of gravity and the cosmological constant problem

Arkani-Hamed N, Georgi H, Schwartz MD (2003) Effective field theory for massive gravitons and gravity in theory space. Ann Phys 305:96–118. doi:10.1016/S0003-4916(03)00068-X

Babichev E, Deffayet C, Ziour R (2009) K-Mouflage gravity. Int J Mod Phys D18:2147–2154. doi:10.1142/S0218271809016107

Bergshoeff EA, Hohm O, Townsend PK (2009) Massive gravity in three dimensions. Phys Rev Lett 102:201301. doi:10.1103/PhysRevLett.102.201301

Boehmer CG, Harko T (2007) Dark energy as a massive vector field. Eur Phys J C 50:423–429. doi:10.1140/epjc/s10052-007-0210-1

Burgess CP (2007) Introduction to effective field theory. Ann Rev Nucl Part Sci 57:329–362. doi:10.1146/annurev.nucl.56.080805.140508

Burrage C, de Rham C, Heisenberg L (2011) de Sitter Galileon. JCAP 1105:025. doi:10.1088/1475-7516/2011/05/025

Burrage C, de Rham C, Seery D, Tolley AJ (2011) Galileon inflation. JCAP 1101:014. doi:10.1088/1475-7516/2011/01/014

Chkareuli G, Pirtskhalava D (2012) Vainshtein mechanism in Λ_3—theories. Phys Lett B 713:99–103. doi:10.1016/j.physletb.2012.05.030

Chow N, Khoury J (2009) Galileon cosmology. Phys Rev D 80:024037. doi:10.1103/PhysRevD.80.024037

Christos C, Ruth G, Nemanja K, Antonio P (2006) DGP Specteroscopy. JHEP 0610:066. doi:10.1088/1126-6708/2006/10/066

Creminelli P, Nicolis A, Trincherini E (2010) Galilean genesis: an alternative to inflation. JCAP 1011:021. doi:10.1088/1475-7516/2010/11/021

Davis SC (2003) Generalized Israel junction conditions for a Gauss-Bonnet brane world. Phys Rev D 67:024030. doi:10.1103/PhysRevD.67.024030

de Fromont P, de Rham C, Heisenberg L, Matas A (2013) Superluminality in the Bi- and Multi-Galileon. JHEP 1307:067. doi:10.1007/JHEP07(2013)067

de Rham C (2010) Massive gravity from Dirichlet boundary conditions. Phys Lett B 688:137–141

de Rham C, Dvali G, Hofmann S, Khoury J, Pujolas O et al (2008) Cascading gravity: extending the Dvali-Gabadadze-Porrati model to higher dimension. Phys Rev Lett 100:251603. doi:10.1103/PhysRevLett.100.251603

de Rham C, Gabadadze G (2010a) Selftuned massive spin-2. Phys Lett B 693:334–338. doi:10.1016/j.physletb.2010.08.043

de Rham C, Gabadadze G (2010b) Generalization of the Fierz-Pauli action. Phys Rev D 82:044020. doi:10.1103/PhysRevD.82.044020 de Rham C, Gabadadze G, Heisenberg L, Pirtskhalava D (2011) Cosmic acceleration and the helicity-0 graviton. Phys Rev D 83:103516. doi:10.1103/PhysRevD.83.103516

de Rham C, Gabadadze G, Heisenberg L, Pirtskhalava D (2012) Non-renormalization and naturalness in a class of scalar-tensor theories. Phys Rev D 87:085017

de Rham C, Gabadadze G, Pirtskhalava D, Tolley AJ, Yavin I (2011) Nonlinear dynamics of 3D massive gravity. JHEP 1106:028. doi:10.1007/JHEP06(2011)028

de Rham C, Gabadadze G, Tolley AJ (2011) Resummation of massive gravity. Phys Rev Lett 106:231101. doi:10.1103/PhysRevLett.106.231101

de Rham C, Heisenberg L (2011) Cosmology of the galileon from massive gravity. Phys Rev D 84:043503. doi:10.1103/PhysRevD.84.043503

de Rham C, Heisenberg L, Ribeiro RH (2015) Quantum corrections in massive gravity at higher loop: mixing matter with gravitons to appear

de Rham C, Tolley AJ (2006) Mimicking Lambda with a spin-two ghost condensate. JCAP 0607:004. doi:10.1088/1475-7516/2006/07/004

de Rham C, Tolley A (2010) DBI and the Galileon reunited. JCAP 1005:015. doi:10.1088/1475-7516/2010/05/015

Deffayet C, Dvali GR, Gabadadze G (2002) Accelerated universe from gravity leaking to extra dimensions. Phys Rev D 65: 044023. doi:10.1103/PhysRevD.65.044023

Deffayet C, Dvali GR, Gabadadze G, Vainshtein AI (2002) Nonperturbative continuity in graviton mass versus perturbative discontinuity. Phys Rev D 65:044026. doi:10.1103/PhysRevD.65.044026

Deffayet C, Esposito-Farese G, Vikman A (2009) Covariant galileon. Phys Rev D 79:084003. doi:10.1103/PhysRevD.79.084003

Deffayet C, Pujolas O, Sawicki I, Vikman A (2010) Imperfect dark energy from kinetic gravity braiding. JCAP 1010:026. doi:10.1088/1475-7516/2010/10/026

Doran M, Lilley MJ, Schwindt J, Wetterich C (2001) Quintessence and the separation of CMB peaks. Astrophys J 559:501–506. doi:10.1086/322253

Dvali GR, Gabadadze G, Porrati M (2000) 4-D gravity on a brane in 5-D Minkowski space. Phys Lett B485:208–214. doi:10.1016/S0370-2693(00)00669-9

Dvali G, Gabadadze G, Shifman M (2002) Diluting cosmological constant via large distance modification of gravity. World Scientific, Singapore, pp 566–581

Dvali G, Gabadadze G, Shifman M (2003) Diluting cosmological constant in infinite volume extra dimensions. Phys Rev D 67:044020. doi:10.1103/PhysRevD.67.044020

Dvali G, Hofmann S, Khoury J (2007) Degravitation of the cosmological constant and graviton width. Phys Rev D 76:084006. doi:10.1103/PhysRevD.76.084006

Fairlie D (2011) Comments on galileons. J Phys A44:305201. doi:10.1088/1751-8113/44/30/305201

Fairlie DB, Govaerts J, Morozov A (1992) Universal field equations with covariant solutions. Nucl Phys B 373:214–232. doi:10.1016/0550-3213(92)90455-K

Fasiello M, Tolley AJ (2013) Cosmological stability bound in massive gravity and bigravity

Fierz M, Pauli W (1939) On relativistic wave equations for particles of arbitrary spin in an electromagnetic field. Proc Roy Soc Lond A173:211–232

Gabadadze G (2009) General relativity with an auxiliary dimension. Phys Lett B 681:89–95

Gabadadze G, Shifman M (2004) Softly massive gravity. Phys Rev D 69:124032. doi:10.1103/PhysRevD.69.124032

Golovnev A, Mukhanov V, Vanchurin V (2008) Vector inflation. JCAP 0806:009. doi:10.1088/1475-7516/2008/06/009

Goon G, Hinterbichler K, Trodden M (2011) Symmetries for Galileons and DBI scalars on curved space. JCAP 1107:017. doi:10.1088/1475-7516/2011/07/017

Hassan SF, Rosen RA (2011) On non-linear actions for massive gravity. JHEP 1107:009. doi:10.1007/JHEP07(2011)009

Hassan SF, Rosen RA (2012) Bimetric gravity from ghost-free massive gravity. JHEP 1202:126. doi:10.1007/JHEP02(2012)126

Hassan SF, Schmidt-May A, von Strauss M (2012) Metric formulation of ghost-free multivielbein theory

Hinterbichler K, Nicolis A, Porrati M (2009) Superluminality in DGP. JHEP 0909:089. doi:10.1088/1126-6708/2009/09/089

Hinterbichler K, Khoury J (2010) Symmetron fields: screening long-range forces through local symmetry restoration. Phys Rev Lett 104:231301. doi:10.1103/PhysRevLett.104.231301

Hinterbichler K, Khoury J (2012) The pseudo-conformal universe: scale invariance from spontaneous breaking of conformal symmetry. JCAP 1204:023. doi:10.1088/1475-7516/2012/04/023

Hinterbichler K, Rosen RA (2012) Interacting spin-2 fields. JHEP 1207:047. doi:10.1007/JHEP07(2012)047

Jiménez JB, Durrer R, Heisenberg L, Thorsrud M (2013) Stability of Horndeski vector-tensor interactions. JCAP 1310:064. doi:10.1088/1475-7516/2013/10/064

Jimenez JB, Froes ALD, Mota DF (2013) Screening vector field modifications of general relativity. Phys Lett B 725:212–217. doi:10.1016/j.physletb.2013.07.032

Khoury J, Weltman A (2004) Chameleon cosmology. Phys Rev D 69:044026. doi:10.1103/PhysRevD.69.044026

Khoury J, Wyman M (2009) N-body simulations of DGP and degravitation theories. Phys Rev D 80:064023. doi:10.1103/PhysRevD.80.064023

Koyama K (2005) Are there ghosts in the self-accelerating brane universe? Phys Rev D 72:123511. doi:10.1103/PhysRevD.72.123511

Koyama K, Niz G, Tasinato G (2011) Analytic solutions in non-linear massive gravity. Phys Rev Lett 107:131101. doi:10.1103/PhysRevLett.107.131101

Luty MA, Porrati M, Rattazzi R (2003) Strong interactions and stability in the DGP model. JHEP 0309:029

Mizuno S, Koyama K (2010) Primordial non-Gaussianity from the DBI Galileons. Phys Rev D 82(103518). doi:10.1103/PhysRevD.82.103518

Nicolis A, Rattazzi R, Trincherini E (2009) The Galileon as a local modification of gravity. Phys Rev D 79:064036. doi:10.1103/PhysRevD.79.064036

Ondo NA, Tolley AJ (2013) Complete decoupling limit of ghost-free massive gravity

Peebles PJE, Ratra B (2003) The cosmological constant and dark energy. Rev Mod Phys 75:559–606. doi:10.1103/RevModPhys.75.559

Scherer S (2003) Introduction to chiral perturbation theory. Adv Nucl Phys 27:277

Siegel W (1986) 16/16 SUPERGRAVITY. Class Quant Grav 3:L47–L48. doi:10.1088/0264-9381/3/2/008

Silva FP, Koyama K (2009) Self-accelerating universe in galileon cosmology. Phys Rev D 80:121301. doi:10.1103/PhysRevD.80.121301

Turner MS, Widrow LM (1988) Inflation-produced, large-scale magnetic fields. Phys Rev D 37: 2743–2754. doi:10.1103/PhysRevD.37.2743. http://link.aps.org/doi/10.1103/PhysRevD.37.2743

Vainshtein AI (1972) To the problem of nonvanishing gravitation mass. Phys Lett B 39:393–394. doi:10.1016/0370-2693(72)90147-5

Weinberg S (2005) The quantum theory of fields, vol 1: Foundations. Cambridge University Press, Cambridge. ISBN 0521670535. http://www.amazon.com/Quantum-Theory-Fields-Foundations/dp/0521670535

Wyman M (2011) Galilean-invariant scalar fields can strengthen gravitational lensing. Phys Rev Lett 106:201102. doi:10.1103/PhysRevLett.106.201102

Part I
Cosmology with Massive Gravity

Chapter 2
Cosmology of Massive Gravity in the Decoupling Limit

In this chapter we will aim the ambitious task of addressing the burdensome problems of cosmology using the framework of massive gravity. More concrete, we will try to answer the questions of whether or not massive gravity can produce a theoretically reliable self-accelerated geometry and whether or not it can also resolve the cosmological constant problem. On these grounds, we will first dare to tackle these problems in a certain approximation of the full theory. This approximation manifest itself in a way such that the helicity ± 2, ± 1, and helicity-0 modes of the massive graviton decouple from each other in the linearized theory, constituting the so-called decoupling limit of massive gravity. The nonlinear self-interactions, and interactions between these modes, are encoded in a few leading higher-dimensional terms in the Lagrangian, as we introduced in detail in Sect. 1.2. Within this approximation we will study two branches of solutions, in which the graviton can

- either form a condensate whose energy density sources self-acceleration,
- or form a condensate whose energy density compensates the cosmological constant.

In the following we will successfully show that it is indeed possible to construct self-accelerated solutions in the decoupling limit of massive gravity. The acceleration is due to a condensate of the helicity-0 field of the massive graviton, which in the decoupling limit is reparametrization invariant as we showed in Sect. 1.2. At this point, it is worth to emphasize again that the helicity-0 field of the massive graviton is not an arbitrary scalar field like in the Quintessence models, since it descends from a full-fledged tensor field. This is the reason why it has no potential interactions, but enters the Lagrangian via very specific derivative interactions fixed by symmetries de Rham and Gabadadze (2010). These derivative interactions induce an effective negative pressure causing the accelerated expansion. We will show that the fluctuations on top of this self-accelerating background are stable.

A rather unexpected result from this branch of self accelerating solutions is enunciated clearly in the fact that from the observational point of view, the obtained self-accelerating background is indistinguishable from that of the Standard Model of cosmology, the ΛCDM model, at leading order. This result was indeed very surprising since we expected that the helicity-0 could have introduced some differences

L. Heisenberg, *Theoretical and Observational Consistency of Massive Gravity*,
Springer Theses, DOI 10.1007/978-3-319-18935-2_2

to the fluctuations and consequently to the evolution of the fluctuations. Essentially, we would have expected that the helicity-0 field could give rise to an additional force at cosmological distance scales modifying the growth and distribution of structure [as it was for instance the case for the helicity-0 mode of the DGP model (Lue et al. 2004; Scoccimarro 2009; Chan and Scoccimarro 2009; Afshordi et al. 2009)], while at shorter scales still being strongly screened via the Vainshtein mechanism (Vainshtein 1972). This would then guarantee the recovery of General Relativity with small departures (Vainshtein 1972; Deffayet et al. 2002), that on the other hand may result in measurable small changes (Dvali et al. 2003; Lue and Starkman 2003) in high-precision Laser Ranging experiments. This is exactly what happens in the self accelerating branch of the DGP model. Anyhow, this does not happen on the self-accelerated background in the massive gravity theory in the decoupling limit. Contrary to expectations, the fluctuation of the helicity-0 field of the massive graviton on top of this self accelerating background decouples in the linearized approximation from an arbitrary source. Consequently, the astrophysical sources will not excite this fluctuation, giving rise exactly to the same ΛCDM results. We should emphasize here that this similarity of the self-accelerated solution and its fluctuations to the ΛCDM results hold in the decoupling limit and it could be that it will not hold beyond this limit.

In addition to the self accelerating branch, we will also show that massive gravity can indeed tackle the cosmological constant problem successfully, avoiding S. Weinberg's no-go theorem (Steven 1989). We will proceed as follows: We will accept a large vacuum energy and show that it gravitates very weakly (Dvali et al. 2002, 2003). The large vacuum energy will not manifest itself as strongly as naively anticipated in General Relativity, i.e. it will be *degravitated*, while all the astrophysical sources will still exhibit the General Relativity behavior (Arkani-Hamed et al. 2002). Strictly speaking, one can think of degravitation as a promotion of Newton's constant to a high pass filter operator thereby modifying the effect of long wavelength sources such as a cosmological constant while recovering General Relativity on shorter wavelengths (Dvali et al. 2002, 2003; Arkani-Hamed et al. 2002). Theories of massive and resonance gravitons are particularly adequate for exhibiting the high pass filter behavior to degravitate the cosmological constant (Dvali et al. 2002, 2003) since they are infra-red modifications of General Raltiviy, meaning that they modify the effects of long wavelength sources. Moreover, it was shown in Dvali et al. (2007) that any causal theory that can degravitate the cosmological constant is a theory of massive gravity or resonance gravitons.

It is important to emphasize that in theories of massive gravity degravitation is a causal process. The real measure of whether or not a source is degravitated is given by its time evolution. During inflation for instance, the vacuum energy driving the acceleration of the Universe will not be degravitated for a long time. It is only after long enough periods of time that the IR modification of gravity kicks in and can effectively slow down an accelerated expansion (Dvali et al. 2002, 2003; Arkani-Hamed et al. 2002). Hence, a crucial ingredient for the degravitation mechanism to work is the existence of a (nearly) static solution in the presence of a cosmological constant towards which the geometry can relax at late time (or after some long period of time). Indeed, Dvali et al. (2007) studied linearized massive gravity demonstrating that in this approximation degravitation takes place after a long enough period of time.

We will successfully show that massive gravity accommodates static solutions while evading any ghost issues at least in the decoupling limit. We will illustrate that in this framework an arbitrary vacuum energy can be neutralized by the effective stress-tensor of the helicity-0 component of the massive graviton. Furthermore, we will put constraints on the two parameters of the theory by demanding the small fluctuations around the degravitating solution to be stable.

An intriguing result we find is that the energy scale at which the interactions of the helicity-0 modes become highly nonlinear is affected by the scale of the degravitated cosmological constant. To be more precise, the interaction scale is higher for larger values of the cosmological constant. Unfortunately, this phenomenon creates a problem by postponing Vainshtein's recovery of General Relativity to shorter and shorter distance scales. As a result, the tests of gravity impose a stringent upper bound on the vacuum energy that can be degravitated in this framework without conflicting measurements of gravity. Disappointingly, this upper bound turns out to be of the same order as the critical energy density of the present-day Universe, $(10^{-3}\ \text{eV})^4$—the value that does not need to be degravitated.

In spite of this low upper bound on the vacuum energy, let us emphasize that there still are two important virtues of the degravitating solution with the low value of the degravitated Cosmological Constant we find here:

- It is a concrete example of how degravitation could work in four-dimensional theories of massive gravity without giving rise to ghost-like instabilities.
- The degravitated solution with small values of cosmological constant can be combined with the self-accelerated solution, to give a satisfactory solution that is in agreement with the existing cosmological and astrophysical data.

Last but not least, the solutions found in the decoupling limit do not necessarily imply the existence of the solutions with identical properties in the full theory. Nevertheless, the decoupling limit solutions should capture the local dynamics at scales well within the present-day Hubble four-volume, as argued in Nicolis et al. (2009). On the other hand, at larger scales the full solutions may be very different from our ones. These differences would kick in at scales comparable to the graviton Compton wavelength. Therefore, our solutions should manifest themselves at least as transients lasting long cosmological times.

As we introduced in detail in Sect. 1.2 the dRGT theory of massive gravity reduces in the decoupling limit to the following interactions for the helicity-2 and helicity-0 components of the massive graviton (de Rham and Gabadadze 2010)

$$\mathcal{L} = -\frac{1}{2} h^{\mu\nu} \mathcal{E}^{\alpha\beta}_{\mu\nu} h_{\alpha\beta} + h^{\mu\nu} \sum_{n=1}^{3} \frac{a_n}{\Lambda_3^{3(n-1)}} X^{(n)}_{\mu\nu}[\Pi], \tag{2.1}$$

where the first term represents the usual kinetic term for the helicity-2 field with $(\mathcal{E}h)_{\mu\nu}$ denoting the linearized Einstein operator acting on $h_{\mu\nu}$ defined in Eq. 1.27,

$a_1 = -1/2$, and $a_{2,3}$ are two arbitrary constants, related to the two parameters from the set $\{c_i, d_i\}$ which characterize a given ghostless theory of massive gravity.

The three symmetric tensors $X^{(n)}_{\mu\nu}[\Pi]$ are composed of the second derivative of the helicity-0 field $\Pi_{\mu\nu} \equiv \partial_\mu \partial_\nu \pi$. In order to maintain reparametrization invariance of the full Lagrangian the tensors $X^{(n)}_{\mu\nu}[\Pi]$ should be identically conserved. These properties uniquely determine the expressions for $X^{(n)}_{\mu\nu}$ at each order of non-linearity. The obtained expressions agree with the results of the direct calculations of de Rham and Gabadadze (2010). A convenient parametrization for the tensors $X^{(n)}_{\mu\nu}$ which we adopt in this chapter is as follows:

$$\begin{aligned} X^{(1)}_{\mu\nu}[\Pi] &= \varepsilon_\mu{}^{\alpha\rho\sigma}\varepsilon_\nu{}^\beta{}_{\rho\sigma}\Pi_{\alpha\beta}, \\ X^{(2)}_{\mu\nu}[\Pi] &= \varepsilon_\mu{}^{\alpha\rho\gamma}\varepsilon_\nu{}^{\beta\sigma}{}_\gamma\Pi_{\alpha\beta}\Pi_{\rho\sigma}, \\ X^{(3)}_{\mu\nu}[\Pi] &= \varepsilon_\mu{}^{\alpha\rho\gamma}\varepsilon_\nu{}^{\beta\sigma\delta}\Pi_{\alpha\beta}\Pi_{\rho\sigma}\Pi_{\gamma\delta}\,. \end{aligned} \tag{2.2}$$

The Lagrangian in the decoupling limit (2.1) represents the *exact* Lagrangian in the sense that it has a finite number of interactions.[1] All the terms higher than quartic order vanish in this limit, making (2.1) a unique theory to which any nonlinear, ghostless extension of massive gravity should reduce in the decoupling limit (de Rham and Gabadadze 2010). Moreover, note that the stress-tensor of external sources only couple to the physical metric $h_{\mu\nu}$. In the basis used in (2.1) there is no direct coupling of π to the stress-tensors. Therefore the Lagrangian (2.1) is invariant with respect to the shifts and the galilean transformations in the internal space of the π field, $\partial_\mu \pi \to \partial_\mu \pi + b_\mu$, where b_μ is a constant four-vector. The latter invariance guarantees that there is no mass nor potential terms generated for π by the loop corrections.

The tree-level coupling of π to the sources arises only after diagonalization: The quadratic mixing $h^{\mu\nu}X^{(1)}_{\mu\nu}$, and the cubic interaction $h^{\mu\nu}X^{(2)}_{\mu\nu}$, can be diagonalized by a nonlinear transformation of $h_{\mu\nu}$, that generates the following coupling of π (de Rham and Gabadadze 2010)

$$\frac{1}{M_{Pl}}\left(-2a_1\eta_{\mu\nu}\pi + \frac{2a_2\partial_\mu\pi\,\partial_\nu\pi}{\Lambda_3^3}\right)T^{\mu\nu}\,. \tag{2.3}$$

Moreover, the above transformation also generates all the Galileon terms for the helicity-0 field, introduced in a different context in Nicolis et al. (2009). In this approximation of massive gravity the coupling of the Galileon field to matter after diagonalization is not only given by $\pi\, T$ as considered in the original Galileon theory (Nicolis et al. 2009), but also includes more generic derivative mixing of the form $\partial_\mu \pi\, \partial_\nu \pi\, T^{\mu\nu}$.

Since the Galileon terms are known to exhibit the Vainshtein recovery of General Relativity at least for static sources (Nicolis et al. 2009), so does the above theory with

[1] Recall that we excluded the helicity-1 part since it only appears quadratically.

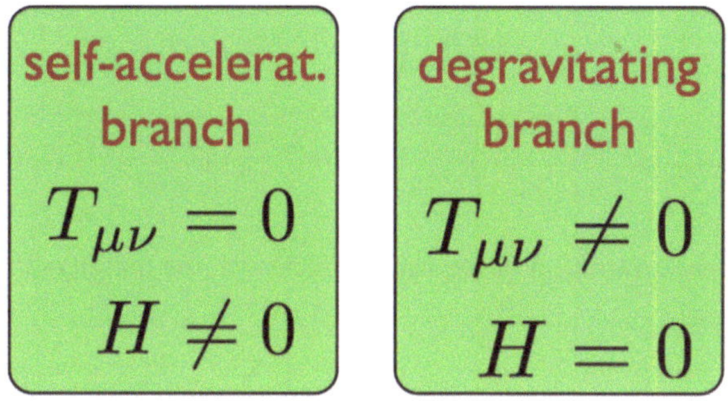

Fig. 2.1 There are two branches of solution: the self-accelerating solution in which the graviton forms a condensate sourcing the acceleration and the degravitating solution where the graviton forms a condensate whose energy density compensates the cosmological constant

$a_3 = 0$. The quartic interaction $h^{\mu\nu}X^{(3)}_{\mu\nu}$, however, cannot be absorbed by any local redefinition of $h_{\mu\nu}$. It is still expected though to admit the Vainshtein mechanism.

However, as we will show in the next section, on the self-accelerated background the fluctuation of the helicity-0 field decouples from an arbitrary source, making the predictions of the theory consistent with General Relativity already in the linearized approximation. This decoupling is a direct consequence of the self-accelerated background and the specific form of the coupling (2.3).

In the following we will explicitly study the two branches of solutions in the decoupling limit: the self-accelerating solution and the degravitating solution (Fig. 2.1).

2.1 The Self-Accelerated Solution in the Decoupling Limit

The universality of the decoupling limit Lagrangian (2.1) for the class of ghostless massive gravities, suggests the possibility of a fairly model-independent phenomenology of the massive theories that should be captured by the limiting Lagrangian (2.1). In the present section, we will be interested in the cosmological solutions in these theories. We will directly work in the decoupling limit, which implies scales much smaller than the Compton wavelength of the graviton. In the case of the self-accelerated de Sitter solution for instance, this corresponds to probing physics within the Hubble scale, which as one would expect, is set by the value of the graviton mass.

2.1.1 Self-Accelerating Background

Below we look for homogeneous and isotropic solutions of the equations of motion that follow from the Lagrangian (2.1). The helicity-0 equation of motion reads as follows:

$$\partial_\alpha\partial_\beta h^{\mu\nu}\left(a_1\varepsilon_\mu{}^{\alpha\rho\sigma}\varepsilon_\nu{}^{\beta}{}_{\rho\sigma}+2\frac{a_2}{\Lambda_3^3}\varepsilon_\mu{}^{\alpha\rho\sigma}\varepsilon_\nu{}^{\beta\gamma}{}_\sigma\Pi_{\rho\gamma}\right.$$
$$\left.+3\frac{a_3}{\Lambda_3^6}\varepsilon_\mu{}^{\alpha\rho\sigma}\varepsilon_\nu{}^{\beta\gamma\delta}\Pi_{\rho\gamma}\Pi_{\sigma\delta}\right)=0\,, \tag{2.4}$$

while variation of the Lagrangian w.r.t. the helicity-2 field gives

$$-\mathcal{E}^{\alpha\beta}_{\mu\nu}h_{\alpha\beta}+\sum_{n=1}^{3}\frac{a_n}{\Lambda_3^{3(n-1)}}X^{(n)}_{\mu\nu}[\Pi]=0. \tag{2.5}$$

We are primarily interested in the self-accelerated solutions of the system (2.4)–(2.5). This solution is obtained by choosing the configuration for π such that the second factor in (2.4) vanishes. This has for consequence to kill the first order mixing between $h_{\mu\nu}$ and π and hence the coupling of π to matter at leading order (which arises after diagonalization of the mixing term). As a consequence the perturbations around the self-accelerated solution we obtain here do not couple to matter. This will be presented in more details in what follows.

For an observer at the origin of the coordinate system, the de Sitter metric can locally (i.e., for times t, and physical distances $|\mathbf{x}|$, much smaller than the Hubble scale H^{-1}) be written as a small perturbation over Minkowski space-time Nicolis et al. (2009)

$$ds^2=[1-\frac{1}{2}H^2x^\alpha x_\alpha]\eta_{\mu\nu}dx^\mu dx^\nu. \tag{2.6}$$

The linearized Einstein tensor for the metric (2.6) is given by

$$G^{\rm lin}_{\mu\nu}=\frac{1}{M_{\rm Pl}}\mathcal{E}^{\alpha\beta}_{\mu\nu}h_{\alpha\beta}=-3H^2\eta_{\mu\nu}. \tag{2.7}$$

For the helicity-0 field we look for the solution of the following isotropic form

$$\pi=\frac{1}{2}q\Lambda_3^3x^\alpha x_\alpha+b\Lambda_3^2t+c\Lambda_3\,, \tag{2.8}$$

where q, b and c are three dimensionless constants.

The equations of motion for the helicity-0 and helicity-2 fields (2.4)–(2.5), therefore, can be recast in the following form

$$H^2\left(-\frac{1}{2}+2a_2q+3a_3q^2\right)=0, \tag{2.9}$$

$$M_{\rm Pl}H^2=2q\Lambda_3^3\left[-\frac{1}{2}+a_2q+a_3q^2\right]. \tag{2.10}$$

Solving the quadratic equation (2.10) for q (for $H \neq 0$), we obtain the Hubble constant of the self-accelerated solution from (2.10). Its magnitude, $H^2 \sim \Lambda_3^3/M_{Pl} = m^2$, is set by the graviton mass, as expected (positivity of H^2 is one of the conditions that we will be demanding below). It is not hard to convince oneself that there exists a whole set of self-accelerated solutions, parametrized by a_2 and a_3. This range, however, will be restricted further by the requirement of stability of the solution, which is the focus of the next section.

2.1.2 Small Perturbations and Stability

Here we investigate the constraints that the requirement of stability imposes on a possible background. Let us adopt a particular solution of the system (2.9)–(2.10) and consider perturbations on the corresponding de Sitter background

$$h_{\mu\nu} = h^b_{\mu\nu} + \chi_{\mu\nu}, \quad \pi = \pi^b + \phi, \tag{2.11}$$

where the superscript b denotes the corresponding background values, and ϕ here stands for the perturbation of the helicity-0 mode. The Lagrangian for the perturbations (up to a total derivative) reads as follows

$$\begin{aligned}\mathcal{L} = &-\frac{1}{2}\chi^{\mu\nu}\mathcal{E}^{\alpha\beta}_{\mu\nu}\chi_{\alpha\beta} + 6(a_2+3a_3q)\frac{H^2M_{Pl}}{\Lambda_3^3}\phi\Box\phi - 3a_3\frac{H^2M_{Pl}}{\Lambda_3^6}(\partial_\mu\phi)^2\Box\phi\\ &+\frac{a_2+3a_3q}{\Lambda_3^3}\chi^{\mu\nu}X^{(2)}_{\mu\nu}[\Phi] + \frac{a_3}{\Lambda_3^6}\chi^{\mu\nu}X^{(3)}_{\mu\nu}[\Phi] + \frac{\chi^{\mu\nu}T_{\mu\nu}}{M_{Pl}},\end{aligned} \tag{2.12}$$

where Φ denotes the four-by-four matrix with the elements $\Phi_{\mu\nu} \equiv \partial_\mu\partial_\nu\phi$. The first term in the first line of the above expression is the Einstein term for $\chi_{\mu\nu}$, the second term is a kinetic term for the scalar, and the third one is the cubic Galileon. The second line contains cubic and quartic interactions between $\chi_{\mu\nu}$ and ϕ, which are identical in form to the corresponding terms in the decoupling limit on Minkowski space-time (2.1). None of these interactions therefore lead to ghost-like instabilities (de Rham and Gabadadze 2010), as long as the ϕ kinetic term is positive definite. Most interestingly, however, there is no quadratic mixing term between χ and ϕ in (2.12), i.e. there is no mixed term like $\chi^{\mu\nu}X^{(1)}_{\mu\nu}$. Since it is only $\chi_{\mu\nu}$ that couples to external sources $T_{\mu\nu}$ in the quadratic approximation, then there will not be a quadratic coupling of ϕ to the sources generated in the absence of the quadratic $\chi - \phi$ mixing. Therefore, for arbitrary external sources, there exist consistent solutions for which the fluctuation of the helicity-0 is not excited, $\phi = 0$. On these solutions one exactly recovers the results of the linearized General Relativity. The above phenomenon provides a mechanism of decoupling the helicity-0 mode from arbitrary external sources! This mechanism is a universal property of the self-accelerating solution in ghostless massive gravity.

Hence, there are no instabilities in (2.12), as long as $a_2 + 3a_3q > 0$. The latter condition, along with the requirement of positivity of H^2, and the equations of motion (2.10), requires that the following system be satisfied:

$$-\frac{1}{2} + 2a_2q + 3a_3q^2 = 0,$$
$$M_{Pl}H^2 = 2q\Lambda_3^3\left[a_2q + a_3q^2 - \frac{1}{2}\right] > 0, \quad a_2 + 3a_3q > 0,$$

for the self-accelerating solution to be physically meaningful. The above system can be solved. The solution is given as follows

$$a_2 < 0, \quad -\frac{2a_2^2}{3} < a_3 < -\frac{a_2^2}{2}, \tag{2.13}$$

while the Hubble constant and q are given by the following expressions

$$H^2 = m^2[2a_2q^2 + 2a_3q^3 - q] > 0, \quad q = -\frac{a_2}{3a_3} + \frac{(2a_2^2 + 3a_3)^{1/2}}{3\sqrt{2}a_3}. \tag{2.14}$$

It is clear from (2.13), that the undiagonalizable interaction $h^{\mu\nu}X^{(3)}_{\mu\nu}$ plays a crucial role for the stability of this class of solutions: All theories without this term (i.e. the ones with $a_3 = 0$) would have ghost-like instabilities on the self-accelerated background.

We therefore conclude that there exists a well-defined class of massive theories with the parameters satisfying the conditions (2.13), which propagate no ghosts on asymptotically flat backgrounds, and also admit stable self-accelerated solutions in the decoupling limit.

As seen from the decoupling limit Lagrangian (2.1), the helicity-0 mode π provides an effective stress-tensor that is felt by the helicity-2 field:

$$T^{\pi}_{\mu\nu} = M_{Pl}\sum_{n=1}^{3}\frac{a_n}{\Lambda_3^{3(n-1)}}X^{(n)}_{\mu\nu}[\Pi]$$
$$= -6qM_{Pl}\Lambda_3^3\left[-\frac{1}{2} + a_2q + a_3q^2\right]\eta_{\mu\nu}. \tag{2.15}$$

It is this stress-tensor that provides the negative pressure density required to drive the acceleration of the Universe. Supplemented by the matter density contribution, it leads to the usual ΛCDM—like cosmological expansion of the background in the sub-horizon approximation used here.

As already mentioned, irrespective of the completion (beyond the Hubble scale) of the self-accelerated solution, it is locally indistinguishable from the ΛCDM model. At horizon scales, however, it is likely that these two scenarios will depart from each other: As we emphasized before, the solutions found in the decoupling limit do not necessarily imply the existence of full solutions with identical properties. A given solution in the decoupling limit can just be a transient state of the full solution. Significant deviations of the latter from the former should kick in at distance/time scales comparable to the graviton Compton wavelength.

2.2 Screening the Cosmological Constant in the Decoupling Limit

One explicit realization of degravitation is expected to occur in massive gravity, where gravity is weaker in the IR, and the graviton mass could play the role of a high-pass filter (Dvali et al. 2002, 2003). In this section we show explicitly how the dRGT theory of massive gravity successfully screens an arbitrarily large cosmological constant in the decoupling limit, while evading any ghost issues and preserving Lorentz invariance.

For convenience we recall the decoupling limit Lagrangian of (2.1) coupled to an external source

$$\mathcal{L} = -\frac{1}{2} h^{\mu\nu} \mathcal{E}^{\alpha\beta}_{\mu\nu} h_{\alpha\beta} + h^{\mu\nu} \sum_{n=1}^{3} \frac{a_n}{\Lambda_3^{3(n-1)}} X^{(n)}_{\mu\nu}[\Pi] + \frac{1}{M_{Pl}} h^{\mu\nu} T_{\mu\nu}\,. \tag{2.16}$$

The equations of motion for the helicity-0 and 2 modes are then

$$-\mathcal{E}^{\alpha\beta}_{\mu\nu} h_{\alpha\beta} + \sum_{n=1}^{3} \frac{a_n}{\Lambda_3^{3(n-1)}} X^{(n)}_{\mu\nu}[\Pi] = -\frac{1}{M_{Pl}} T_{\mu\nu}\,, \tag{2.17}$$

and

$$\begin{aligned}
&\left(a_1 + \frac{a_2}{\Lambda_3^3} \Box \pi + \frac{3a_3}{2\Lambda_3^6} \left([\Pi]^2 - [\Pi^2]\right)\right) \left[\Box h - \partial_\alpha \partial_\beta h^{\alpha\beta}\right] \\
&+ \frac{1}{\Lambda_3^3} \left(a_2 \Pi_{\mu\nu} - 3\frac{a_3}{\Lambda_3^3} \left(\Pi^2_{\mu\nu} - \Box\pi\, \Pi_{\mu\nu}\right)\right) \left[2\partial^\mu \partial_\alpha h^{\alpha\nu} - \Box h^{\mu\nu} - \partial^\mu \partial^\nu h\right] \\
&- \frac{3a_3}{\Lambda_3^6} \left(\Pi_{\mu\alpha} \Pi_{\nu\beta} - \Pi_{\mu\nu} \Pi_{\alpha\beta}\right) \partial^\alpha \partial^\beta h^{\mu\nu} = 0\,.
\end{aligned} \tag{2.18}$$

We now focus on a pure cosmological constant source, $T_{\mu\nu} = -\lambda \eta_{\mu\nu}$, and make use of the following ansatz,

$$h_{\mu\nu} = -\frac{1}{2}H^2x^2M_{Pl}\,\eta_{\mu\nu}\,, \tag{2.19}$$

$$\pi = \frac{1}{2}q\,x^2\Lambda_3^3\,. \tag{2.20}$$

The equations of motion then simplify to

$$\left(-\frac{1}{2}M_{Pl}H^2 + \sum_{n=1}^{3} a_n\,q^n\Lambda_3^3\right)\eta_{\mu\nu} = -\frac{\lambda}{6M_{Pl}}\eta_{\mu\nu}\,, \tag{2.21}$$

$$H^2\left(a_1 + 2a_2q + 3a_3q^2\right) = 0\,. \tag{2.22}$$

As we will see below, this system of equations admits two branches of solutions, a degravitating one, for which the geometry remains flat (mimicking the late-time part of the relaxation process), and a de Sitter branch which is closely related to the standard GR de Sitter solution. We start with the degravitating branch before exploring the more usual de Sitter solution and show that the stability of these branches depends on the free parameters $a_{2,3}$, as well as the magnitude of the cosmological constant.

2.2.1 The Degravitating Branch

It is easy to check that the geometry can remain flat i.e. $H = 0$ and $g_{\mu\nu} \equiv \eta_{\mu\nu}$, despite the presence of the cosmological constant. Such solutions are possible due to the presence of the extra helicity-0 mode that carries the source instead of the usual metric. With $H = 0$, Eq. (2.22) is trivially satisfied, while the modified Einstein equation (2.21) determines the coefficient (which we denote by q_0 here) for the helicity-0 field in (2.20),

$$a_1q_0 + a_2q_0^2 + a_3q_0^3 = -\frac{\tilde{\lambda}}{6}\,, \tag{2.23}$$

in terms of the dimensionless quantity $\tilde{\lambda} = \lambda/\Lambda_3^3M_{Pl}$. Notice that as long as the parameter a_3 is present, Eq. (2.23) has always at least one real root. There is therefore a flat solution for arbitrarily large cosmological constant.

Let us now briefly comment on the stability of the flat solution, as this has important consequences for the relaxation mechanism behind degravitation. We consider the field fluctuations above the static solution,

$$\pi = \frac{1}{2}q_0\Lambda_3^3x^2 - \phi/\kappa, \tag{2.24}$$

$$T_{\mu\nu} = -\lambda\eta_{\mu\nu} + \tau_{\mu\nu}, \tag{2.25}$$

where q_0 is related to λ via (2.23) and the coupling κ is determined by

$$\kappa = 2(a_1 + 2a_2 q_0 + 3a_3 q_0^2)\,. \tag{2.26}$$

To the leading order, the action for these fluctuations is then simply given by

$$\mathcal{L}^{(2)} = -\frac{1}{2} h^{\mu\nu} \mathcal{E}^{\alpha\beta}_{\mu\nu} h_{\alpha\beta} - \frac{1}{2} h^{\mu\nu} X^{(1)}_{\mu\nu}[\Phi] + \frac{1}{M_{Pl}} h^{\mu\nu} \tau_{\mu\nu}\,, \tag{2.27}$$

with $\Phi_{\mu\nu} = \partial_\mu \partial_\nu \phi$. The stability of this theory is better understood when working in the Einstein frame where the helicity-0 and -2 modes decouple. This is achieved by performing the change of variable,

$$h_{\mu\nu} = \bar{h}_{\mu\nu} + \phi\, \eta_{\mu\nu}\,, \tag{2.28}$$

which brings the action to the following form

$$\mathcal{L}^{(2)} = -\frac{1}{2} \bar{h}^{\mu\nu} \mathcal{E}^{\alpha\beta}_{\mu\nu} \bar{h}_{\alpha\beta} + \frac{3}{2}\, \phi \Box \phi + \frac{1}{M_{Pl}} \left(\bar{h}^{\mu\nu} + \phi\, \eta^{\mu\nu} \right) \tau_{\mu\nu}. \tag{2.29}$$

Stability of the static solution is therefore manifest for any region of the parameter space for which κ is real and does not vanish. As already mentioned, if $a_3 \neq 0$ there is always a real solution to (2.23), which is therefore stable for $\kappa \neq 0$. This suggests the presence of a flat late-time attractor solution for degravitation. The special case $a_3 = 0$ is discussed separately below.

2.2.2 de Sitter Branch

In the presence of a cosmological constant, the field equations (2.21) and (2.22) also admit a second branch of solutions; these connect with the self-accelerating branch presented in Sect. 2.1, and we refer to them as the de Sitter solutions. The parameters for these solutions should satisfy

$$a_1 + 2a_2 q_{dS} + 3a_3 q_{dS}^2 = 0\,, \tag{2.30}$$

$$H_{dS}^2 = \frac{\lambda}{3M_{Pl}^2} + \frac{2\Lambda_3^3}{M_{Pl}} \left(a_1 q_{dS} + a_2 q_{dS}^2 + a_3 q_{dS}^3 \right)\,. \tag{2.31}$$

This solution is closer to the usual GR de Sitter configuration and only exists if $a_2^2 \geq 3a_1 a_3$. The stability of this solution can be analyzed as previously by looking at fluctuations around this background configuration,

$$\pi = \frac{1}{2} q_{dS} \Lambda_3^3 x^2 + \phi, \tag{2.32}$$

$$h_{\mu\nu} = -\frac{1}{2} H_{dS}^2 x^2 \eta_{\mu\nu} + \chi_{\mu\nu}, \tag{2.33}$$

$$T_{\mu\nu} = -\lambda \eta_{\mu\nu} + \tau_{\mu\nu}. \tag{2.34}$$

To second order in fluctuations, the resulting action is then of the form

$$\mathcal{L}^{(2)} = -\frac{1}{2}\chi^{\mu\nu}\mathcal{E}_{\mu\nu}^{\alpha\beta}\chi_{\alpha\beta} + \frac{6H_{dS}^2 M_{Pl}}{\Lambda_3^3}(a_2 + 3a_3 q_{dS})\,\phi\Box\phi + \frac{1}{M_{Pl}}\chi^{\mu\nu}\tau_{\mu\nu}. \tag{2.35}$$

It is interesting to point out again that the helicity-0 fluctuation ϕ then decouples from matter sources at quadratic order (however the coupling reappears at the cubic order). Stability of this solution is therefore ensured if the parameters satisfy one of the following three constraints, (setting $a_1 = -1/2$ and $\tilde{\lambda} > 0$)

$$a_2 < 0 \quad \text{and} \quad -\frac{2a_2^2}{3} \leq a_3 < \frac{1 - 3a_2\tilde{\lambda} - (1 - 2a_2\tilde{\lambda})^{3/2}}{3\tilde{\lambda}^2}, \tag{2.36}$$

or

$$a_2 < \frac{1}{2\tilde{\lambda}} \quad \text{and} \quad a_3 > \frac{1 - 3a_2\tilde{\lambda} + (1 - 2a_2\tilde{\lambda})^{3/2}}{3\tilde{\lambda}^2}, \tag{2.37}$$

or

$$a_2 \geq \frac{1}{2\tilde{\lambda}} \quad \text{and} \quad a_3 > -\frac{2}{3}a_2^2. \tag{2.38}$$

These are consistent with the results (2.13) found for the self-accelerating solution in the absence of a cosmological constant. Notice here that in the presence of a cosmological constant, the accelerating solution can be stable even when $a_3 = 0$. This branch of solutions therefore connects with the usual de Sitter one of GR.

2.2.3 Diagonalizable Action

In Sect. 2.1 we have emphasized the importance of the contribution of $X_{\mu\nu}^{(3)}$ for the stability of the self-accelerating solution. However, in the presence of a nonzero cosmological constant, this contribution is not a priori essential for the stability of either the degravitating or the de Sitter branches. Furthermore, since the helicity-0 and -2 modes can be diagonalized at the nonlinear level when $a_3 = 0$, as was explicitly shown in de Rham and Gabadadze (2010), we will study this special case separately below. In particular, we will show that it leads to certain special bounds both in the degravitating and de Sitter branches of solution.

Stability: To start with, when $a_3 = 0$, the degravitating solution only exists if

$$2a_2\tilde{\lambda} < 3a_1^2 \,. \tag{2.39}$$

This bound ensures the absence of ghost-like instabilities around the degravitating solution. Assuming that the parameters $a_{1,2} = \mathcal{O}(1)$ take some natural values then the situation $a_2 > 0$ implies a severe constraint on the value of the vacuum energy that can be degravitated. This is similar to the bound in the non-linear realization of massive gravity (de Rham 2010), as well as in codimension-two deficit angle solutions, $\lambda \lesssim m^2 M_{Pl}^2$. The situation $a_2 < 0$ on the other hand allows for an arbitrarily large cosmological constant.

On the other hand, the bound $a_2^2 \geq 3a_1 a_3$ for the existence of the de Sitter solution is always satisfied if $a_3 = 0$. However, the constraints on the parameters (2.36)–(2.38) which guarantee the absence of ghosts on the de Sitter branch imply that

$$2a_2\tilde{\lambda} > 3a_1^2 \,. \tag{2.40}$$

In this specific case then, we infer that when the Sitter solution is stable, the degravitating branch does not exist, and when the degravitating branch exists the de Sitter solution is unstable. Therefore, at each point in the parameter space there is only one, out of these two solutions, that makes sense. In the more general case where $a_3 \neq 0$ the situation is however much more subtle and it might be possible to find parameters for which both branches exist and are stable simultaneously.

Einstein's frame: Let us now work instead in the Einstein frame, where the helicity-2 and -0 modes are diagonalized (which is possible as long as $a_3 = 0$). The transition to Einstein's frame is performed by the change of variable (de Rham and Gabadadze 2010)

$$h_{\mu\nu} = \bar{h}_{\mu\nu} - 2a_1 \pi\, \eta_{\mu\nu} + \frac{2a_2}{\Lambda_3^3} \partial_\mu \pi\, \partial_\nu \pi \,, \tag{2.41}$$

such that the action takes the form

$$\begin{aligned} \mathcal{L} = &-\frac{1}{2}\bar{h}^{\mu\nu}(\mathcal{E}\bar{h})_{\mu\nu} + 6a_1^2\, \pi \,\Box\, \pi - \frac{6a_2 a_1}{\Lambda_3^3}(\partial\pi)^2[\Pi] \\ &+\frac{2a_2^2}{\Lambda_3^6}(\partial\pi)^2\left([\Pi^2] - [\Pi]^2\right) \\ &+\frac{1}{M_{Pl}}\left(\bar{h}_{\mu\nu} - 2a_1\, \pi\, \eta_{\mu\nu} + \frac{2a_2}{\Lambda_3^3}\partial_\mu \pi\, \partial_\nu \pi\right)T^{\mu\nu}, \end{aligned} \tag{2.42}$$

and the structure of the Galileon becomes manifest. Notice however, that the coefficients of the different Galileon interactions are not arbitrary. Furthermore, the coupling to matter includes terms of the form $\partial_\mu \pi\, \partial_\nu \pi\, T^{\mu\nu}$, absent in the original

Galileon formalism (Nicolis et al. 2009). Both of these distinctions play a crucial role in screening the cosmological constant—the task which was thought impossible in the original Galileon theory. Here, however, as long as the bound (2.39) is satisfied, the solution for π reads

$$\pi = \frac{1}{2} q_0 \Lambda_3^3 x^2 \qquad \text{with} \qquad a_1 q_0 + a_2 q_0^2 = -\frac{\tilde{\lambda}}{6}, \tag{2.43}$$

while the helicity-2 mode $\bar{h}_{\mu\nu}$ now takes the form

$$\bar{h}_{\mu\nu} = \left(\frac{\xi}{2} - \frac{\lambda}{6M_{Pl}}\right) x^2 \eta_{\mu\nu} + \xi\, x_\mu x_\nu, \tag{2.44}$$

with ξ being an arbitrary gauge freedom parameter. Fixing $\xi = -2a_2 q_0^2 \Lambda_3^3$, the physical metric is then manifestly flat:

$$\begin{aligned} g_{\mu\nu} &= \eta_{\mu\nu} + \frac{1}{M_{Pl}}\left(\bar{h}_{\mu\nu} - 2a_1\, \pi\, \eta_{\mu\nu} + \frac{2a_2}{\Lambda_3^3} \partial_\mu \pi\, \partial_\nu \pi\right) \\ &= \eta_{\mu\nu} - \frac{\lambda_3^3}{M_{Pl}}\left(a_1 q_0 + a_2 q_0^2 + \frac{\tilde{\lambda}}{6}\right) x^2 \eta_{\mu\nu} + \frac{x_\mu x_\nu}{M_{Pl}}\left(\xi + 2a_2 q_0^2 \Lambda_3^3\right) \\ &\equiv \eta_{\mu\nu}. \end{aligned}$$

To reiterate, the specific nonlinear coupling to matter that naturally arises in the ghostless theory of massive gravity is essential for the screening mechanism to work. This allows us to understand why neither DGP nor an ordinary Galileon theory are capable of achieving degravitation. As we already pointed out, the Galileon interactions arise naturally after diagonalization. However, let us summarize the common and different points between Galileon theory and the interactions in the decoupling limit after diagonalization.

Common	Differences
IR modification of gravity as due to a light scalar field with non-linear derivative interactions ($\to$ Vainshtein mechanism)	Non-diagonazable interaction $\frac{a_3}{\Lambda_3^6} h^{\mu\nu} X^{(3)}_{\mu\nu}$ which is important for the self-accelerating solution
Respects the symmetry $\pi \to \pi + c + b_\mu x^\mu$	Extra coupling $\partial_\mu \pi\, \partial_\nu \pi\, T^{\mu\nu}$ which is important for the degravitating solution
Second order equations of motion, containing at most two time derivatives	Only two free parameters
Non-renormalization theorem applies	Observational differences

2.2.4 Phenomenology

Let us now focus on the phenomenology of the degravitating solution. This mechanism relies crucially on the extra helicity-0 mode in the massive graviton. However tests of gravity severely constrain the presence of additional scalar degrees of freedom. As is well known in theories of massive gravity, the helicity-0 mode can evade fifth force constraints in the vicinity of matter if the helicity-0 mode interactions are important enough to freeze out the field fluctuations, Vainshtein (1972).

Around the degravitating solution, the scale for helicity-0 interactions are no longer governed by the parameter Λ_3, but rather by the scale determined by the cosmological constant $\tilde{\Lambda}_3 \sim (\lambda/M_{Pl})^{1/3}$. To see this, let us pursue the analysis of the fluctuations around the degravitating branch (2.24) and keep the higher order interactions. The resulting Lagrangian is then

$$\mathcal{L}^{(2)} = -\frac{1}{2}h^{\mu\nu}\mathcal{E}^{\alpha\beta}_{\mu\nu}h_{\alpha\beta} - \frac{1}{2}h^{\mu\nu}\left(X^{(1)}_{\mu\nu}[\Phi] + \frac{\tilde{a}_2}{\tilde{\Lambda}^3}X^{(2)}_{\mu\nu}[\Phi] + \frac{\tilde{a}_3}{\tilde{\Lambda}^6}X^{(3)}_{\mu\nu}[\Phi]\right) + \frac{1}{M_{Pl}}h^{\mu\nu}\tau_{\mu\nu}, \tag{2.45}$$

with

$$\frac{\tilde{a}_2}{\tilde{\Lambda}^3} = -2\frac{a_2 + 3a_3q_0}{\Lambda_3^3\kappa^2}, \quad \text{and} \quad \frac{\tilde{a}_3}{\tilde{\Lambda}^6} = -\frac{2a_3}{\Lambda_3^6\kappa^3}. \tag{2.46}$$

Assuming $a_{2,3} \sim \mathcal{O}(1)$, a large cosmological constant $\tilde{\lambda} \gg 1$, implies $q_0 \gg 1$, so that $a_3q_0^2 \gg a_2q_0 \gg a_1$ and $\kappa \sim a_3q_0$ such that

$$\frac{\tilde{a}_2}{\tilde{\Lambda}^3} \sim \frac{1}{\Lambda_3^3a_3q_0^3} \sim \frac{1}{\Lambda_3^3\tilde{\lambda}} \sim \frac{M_{Pl}}{\lambda} \tag{2.47}$$

(notice that this result is maintained even if $a_3 = 0$), and similarly

$$\frac{\tilde{a}_3}{\tilde{\Lambda}^6} \sim \left(\frac{M_{Pl}}{\lambda}\right)^2. \tag{2.48}$$

To evade fifth force constrains within the solar system, the scale $\tilde{\Lambda}$ should therefore be small enough to allow for the nonlinear interactions to dominate over the quadratic contribution and enable the Vainshtein mechanism. In the DGP model this typically imposes the constraint, $\tilde{\Lambda}^3/M_{Pl} \lesssim (10^{-33}\text{ eV})^2$, while this value can be pushed by a few orders of magnitude in the presence of Galileon interactions, (Nicolis et al. 2009; Burrage and Seery 2010). Therefore, the allowed value of vacuum energy that can be screened without being in conflict with observations is fairly low, of the order of $(10^{-3}\text{ eV})^4$ or so.

An alternative would be to impose a hierarchy between the dimensionless coefficients a_i. Since the Galilean interactions satisfy a non-renormalization theorem (Nicolis and Rattazzi 2004), which we will discuss in more detail in Chap. 5 such a tuning would remain technically natural. To explore this avenue in a simple way, let us set $a_3 = 0$. In that case, the effective strong coupling scale is given by

$$\tilde{\Lambda}^3 = \Lambda_3^3 \frac{\frac{3}{4} - 2a_2\tilde{\lambda}}{a_2}. \tag{2.49}$$

The strong coupling scale can then be tuned to small values by adjusting the parameter a_2 within the very small window

$$|a_2\tilde{\lambda} - \frac{3}{8}| \lesssim \frac{(10^{-33}\text{eV})^2 M_{Pl}}{\Lambda_3^3}. \tag{2.50}$$

Therefore even when allowing a hierarchy between the parameters, once they are fixed only very restricted values of the degravitated cosmological constant would be compatible with solar system tests. The previous argument would have been unaffected if we had set $a_3 \neq 0$.

The above constraint on the vacuum energy that can be degravitated makes the present framework not viable phenomenologically for solving the old cosmological constant problem. There may be a way out of this setback though: As mentioned previously, one may envisage a cosmological scenario in which the neutralization of vacuum energy takes place before the Universe enters the epoch for which the Vainshtein mechanism is absolutely necessary to suppress the helicity-0 fluctuations. Such an epoch should certainly be before the radiation domination. During that epoch, however, the cosmological evolution should reset itself—perhaps via some sort of phase transition—to continue subsequent evolution along the other branch of the solutions that exhibits the standard early behavior followed by the self-acceleration. This scheme would have to address the cosmological instabilities discussed in Grisa and Sorbo (2010), Berkhahn et al. (2010). Moreover, the viability of such a scenario would depend on properties of the degravitating solution in the full theory—which are not known. Therefore, we do not rely on this possibility.

Nevertheless, there are certain important virtues to the degravitating solution with the low value of the degravitated cosmological constant. This is an example of high importance in understanding how S. Weinberg's no-go theorem can be evaded in principle. As already emphasized in Rham et al. (2008, 2009), such mechanisms evade the no-go theorem by employing a field which explicitly breaks Poincaré invariance in its vacuum configuration $\pi \sim x^2$, while keeping the physics insensitive to this breaking. Indeed, physical observables are only sensitive to $\Pi_{\mu\nu} = \partial_\mu\partial_\nu \pi$ which is clearly Poincaré invariant, while the configuration of the π field itself has no direct physical bearing. This is built in the specific Galileon symmetry of the theory, and is a consequence of the fact that π is not an arbitrary scalar field but rather descends as the helicity-0 mode of the massive graviton. More precisely, under a

Poincaré transformation, $x^\mu \rightarrow \Lambda^\mu_\nu x^\nu + a^\mu$, the configuration for π transforms as $x^2 \rightarrow x^2 + v_\mu x^\mu + c$, with $v_\mu = 2a_\nu \Lambda^\nu_\mu$ and $c = a^2$ which is precisely the Galileon transformation for π under which the action is invariant. In other words the Poincaré symmetry is still realized up to a Galilean transformation (or, there is a diagonal subgroup of Poincaré and internal "galilean" groups that remains unbroken by the VEV of the π field).

Thus, we have presented here the crucial steps towards a non-linear realization of degravitation within the context of massive gravity, and this, without introducing any ghosts (at least in the decoupling limit). The arguments presented here only rely on the decoupling limit and it is reasonable to doubt their validity beyond that regime.

2.3 Summary and Critics

In this chapter we have addressed the potential impact of massive gravity on cosmology. We studied the self-accelerating and degravitating solutions in the decoupling limit of massive gravity and put constraints on the two free parameters of the theory from instability conditions in the cosmological evolution demanding the absence of ghost and Laplacian instabilities. We have shown that massive gravity can be used to construct self accelerating solutions in the decoupling limit. The helicity-0 degree of freedom of the massive graviton forms a condensate whose energy density sources self-acceleration and small fluctuations around these self-accelerating solutions are stable. Furthermore, the fluctuations of the helicity-0 field do not couple to the fluctuations of the helicity-2 field and hence to the matter fields such that the cosmological evolution is exactly as in the standard ΛCDM model. We have also demonstrated that massive gravity can screen an arbitrarily large cosmological constant in the decoupling limit without giving rise to any ghost. Unfortunately, the allowed value of the vacuum energy that can be screened without being in conflict with observations is fairly low. This conflict with the Vainshtein mechanism renders the degravitation solution as found here phenomenologically not viable for solving the old cosmological constant problem. Nevertheless, it is the first time that an explicit model can present a way out from Weinberg's no-go theroem. A possible way out of the conflict between the General Relativity recovery scale and the large Cosmological Constant could be to envisage a cosmological scenario in which degravitation of the vacuum energy takes place before the Universe enters the radiation dominated epoch—say during the inflationary period, or even earlier. By the end of that epoch then the cosmology should reset itself to continue evolution along the other branch of the solutions that exhibits the standard early behavior followed by the self-acceleration. The existence of such a transition would depend on properties of the degravitating solution in the full theory.

As already mentioned before, these solutions found in the decoupling limit do not guaranty the existence of full solutions with identical properties in the full theory. The solutions in the decoupling limit could be considered just as a transient state of the full solution. Since our work, there has been made a quite a lot of progress in studying the self-accelerating solutions, nevertheless the found solutions in the literature seem to be plagued by instabilities. On the degravitating solutions side, there has been little progress in the full theory and these solutions have been left aside so far in the literature. Even if the decoupling limit of massive gravity fails to degravitate an arbitrary large cosmological constant in order to make the Vainshtein mechanism work, it could be that in the full theory there exists a cosmological scenario in which the degravitation of the vacuum energy is not in conflict with the Vainshtein mechanism. This is worth the effort to investigate in the future.

We would like now take a critical viewpoint on the analysis performed in this chapter to discuss the limitations. The first limitation is the fact that the solutions found here are only valid in the approximation we made, on scales smaller than the Hubble scale. The second more worrisome limitation is the negligence of the helicity-1 field. As we emphasized before, the helicity-1 field enters only quadratically, or in higher order terms in the Lagrangian, and therefore we set it to be zero.

Since the appearance of our work there has been a flurry of investigations related to the self- accelerating solutions of the full theory in the literature, which go beyond the study represented here in this thesis. It has been shown that if the fiducial metric is chosen to be flat than for the physical metric there is no flat FLRW solutions (D'Amico et al. 2011) in the full theory beyond the decoupling limit. One way out of this no-go solution is to make the Stueckelberg field carry anisotropies but still keep the geometry FLRW. Nevertheless, solutions found in this way turn out to have strongly coupled degrees of freedom, meaning that some of the degrees of freedom lose their kinetic terms. Alternatively, one can accept non-FLRW solutions but puts constraints on the magnitude of the mass of the graviton coming from the consistency with known constraints on homogeneity and isotropy. This would rely on the successful implementation of the Vainshtein mechanism in the cosmological evolution which so far has not been investigated in detail. Even if flat FLRW has been proven not to exist, there are successful constructions of open FLRW solutions (Gumrukcuoglu et al. 2011). However, at the level of non-linear perturbations instabilities pop up again and render these solutions phenomenologically not viable. These negative outcomes forced the considerations of more general fiducial metric. Indeed, if one assumes a de Sitter reference metric, then one can find FLRW solutions. But the de Sitter reference metric brings other problems along. The Higuchi bound imposes the mass of the graviton to be $m^2 > H^2$ which turns these solutions inconsistent with the observational constraints. Similarly, the generalization of the fiducial metric to a FLRW metric forces a generalized Higuchi bound and one encounters similar problems (Fasiello and Tolley 2012). As we already explained in the introduction the dRGT theory is based on a framework in which the massive graviton propagates on top of a fixed background reference metric. The generalization of dRGT gave rise to theories of ghost-free bimetric gravity in which the reference metric becomes dynamical as well. This bimetric generalization of dRGT gives rise to stable self-accelerating solutions without imposing problems related to the Higuchi bound and so put into operation new exciting research directions (Fasiello and Tolley 2013). There has also been other extensions of massive gravity, like time-depending mass

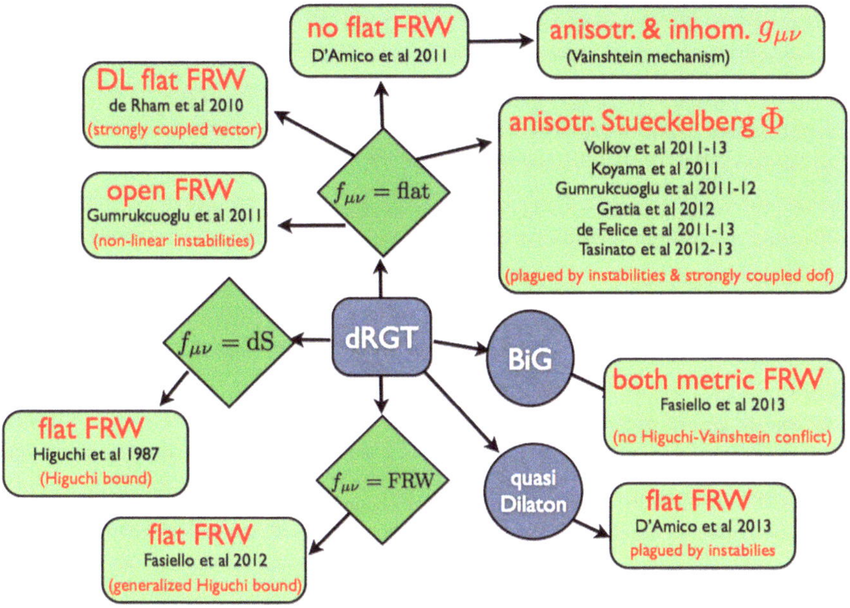

Fig. 2.2 The cosmological "tree" of massive gravity

of the graviton (Huang 2012) or adding additional degrees of freedom [Quasi-dilaton D'Amico et al. (2013)] from which some of the generalization might yield stable self-accelerating solutions (Fig. 2.2).

References

Afshordi N, Geshnizjani G, Khoury J (2009) Do observations offer evidence for cosmological-scale extra dimensions? JCAP 0908:030. doi:10.1088/1475-7516/2009/08/030

D'Amico G, Gabadadze G, Hui L, Pirtskhalava D (2013) Quasidilaton: theory and cosmology. Phys Rev D 87:064037. doi:10.1103/PhysRevD.87.064037. http://link.aps.org/doi/10.1103/PhysRevD.87.064037

Arkani-Hamed N, Dimopoulos S, Dvali G, Gabadadze G (2002) Nonlocal modification of gravity and the cosmological constant problem

Berkhahn F, Dietrich DD, Hofmann S (2010) Self-protection of massive cosmological gravitons. JCAP 1011:018. doi:10.1088/1475-7516/2010/11/018

Burrage C, Seery D (2010) Revisiting fifth forces in the Galileon model. JCAP 1008:011. doi:10.1088/1475-7516/2010/08/011

Chan KC, Scoccimarro R (2009) Large-scale structure in brane-induced gravity II. Numerical simulations. Phys Rev D 80:104005. doi:10.1103/PhysRevD.80.104005

D'Amico G, de Rham C, Dubovsky S, Gabadadze G, Pirtskhalava D, Tolley AJ (2011) Massive cosmologies. Phys Rev D 84:124046. doi:10.1103/PhysRevD.84.124046. http://link.aps.org/doi/10.1103/PhysRevD.84.124046

de Rham C (2010) Massive gravity from Dirichlet boundary conditions. Phys Lett B 688:137–141

de Rham C, Dvali G, Hofmann S, Khoury J, Pujolas O et al (2008) Cascading gravity: extending the Dvali-Gabadadze-Porrati model to higher dimension. Phys Rev Lett 100:251603. doi:10.1103/PhysRevLett.100.251603
de Rham C, Khoury J, Tolley AJ (2009) Flat 3-brane with tension in cascading gravity. Phys Rev Lett 103:161601. doi:10.1103/PhysRevLett.103.161601
de Rham C, Gabadadze G (2010) Generalization of the fierz-pauli action. Phys Rev D 82:044020. doi:10.1103/PhysRevD.82.044020
Deffayet C, Dvali GR, Gabadadze G, Vainshtein AI (2002) Nonperturbative continuity in graviton mass versus perturbative discontinuity. Phys Rev D 65:044026. doi:10.1103/PhysRevD.65.044026
Dvali G, Gruzinov A, Zaldarriaga M (2003) The accelerated universe and the moon. Phys Rev D 68:024012. doi:10.1103/PhysRevD.68.024012
Dvali G, Hofmann S, Khoury J (2007) Degravitation of the cosmological constant and graviton width. Phys Rev D 76:084006. doi:10.1103/PhysRevD.76.084006
Dvali G, Gabadadze G, Shifman M (2002) Diluting cosmological constant via large distance modification of gravity. pp 566–581
Dvali G, Gabadadze G, Shifman M (2003) Diluting cosmological constant in infinite volume extra dimensions. Phys Rev D 67:044020. doi:10.1103/PhysRevD.67.044020
Fasiello M, Tolley AJ (2013) Cosmological stability bound in massive gravity and bigravity
Fasiello M, Tolley AJ (2012) Cosmological perturbations in massive gravity and the Higuchi bound. JCAP 1211:035. doi:10.1088/1475-7516/2012/11/035
Grisa L, Sorbo L (2010) Pauli-Fierz gravitons on Friedmann-Robertson-Walker background. Phys Lett B 686:273–278. doi:10.1016/j.physletb.2010.02.072
Gumrukcuoglu AE, Lin C, Mukohyama S (2011) Open FRW universes and self-acceleration from nonlinear massive gravity. JCAP 1111:030. doi:10.1088/1475-7516/2011/11/030
Huang Q-G, Piao Y-S, Zhou S-Y (2012) Mass-varying massive gravity. Phys Rev D 86:124014. doi:10.1103/PhysRevD.86.124014
Lue A, Scoccimarro R, Starkman GD (2004) Probing Newton's constant on vast scales: DGP gravity, cosmic acceleration and large scale structure. Phys Rev D 69:124015. doi:10.1103/PhysRevD.69.124015
Lue A, Starkman G (2003) Gravitational leakage into extra dimensions: probing dark energy using local gravity. Phys Rev D 67:064002. doi:10.1103/PhysRevD.67.064002
Nicolis A, Rattazzi R, Trincherini E (2009) The Galileon as a local modification of gravity. Phys Rev D 79:064036. doi:10.1103/PhysRevD.79.064036
Nicolis A, Rattazzi R (2004) Classical and quantum consistency of the DGP model. JHEP 0406:059. doi:10.1088/1126-6708/2004/06/059
Scoccimarro R (2009) Large-scale structure in brane-induced gravity I. Perturbation Theory Phys Rev D 80:104006. doi:10.1103/PhysRevD.80.104006
Steven W (1989) The cosmological constant problem. Rev Mod Phys 61:1–23
Vainshtein AI (1972) To the problem of nonvanishing gravitation mass. Phys Lett B 39:393–394. doi:10.1016/0370-2693(72)90147-5

Chapter 3
Proxy Theory

The exact non-linear theory of massive gravity has a very complex structure and therefore in this section we will take an alternative approach of covariantizing the Lagrangian in the decoupling limit, and use the resulting theory as a proxy and study the cosmology in this proxy theory. This procedure was successfully used in the context of DGP model and there it was shown to be a convenient tool to analyze the cosmology and perturbations. Once we covariantize the decoupling limit Lagrangian, the resulting proxy theory is not a massive gravity theory anymore but rather a non-minimally coupled scalar-tensor theory. Nevertheless, they might have similar cosmological properties since they share the same decoupling limit. In this section we first show how to construct the proxy theory from the decoupling limit and illustrate its theoretical consistency. We compute the covariant equations of motions and prove that they contain at most second order derivatives acting on the fields. We then work out the consequences of this covariantization for cosmology and specifically for late-time acceleration. We investigate the stability conditions for perturbations in detail and comment on the existing of degravitating solutions. We then move onto more general cosmology and follow the helicity-0 mode contribution to the Universe throughout its evolution and its effects on structure formation. This is based on our work presented in de Rham and Heisenberg (2011).

For the construction of our proxy theory let us start with the first interaction between the helicity-0 and helicity-2 modes in the decoupling limit and integrate it by part

$$h^{\mu\nu}X^{(1)}_{\mu\nu} = h^{\mu\nu}(\partial_\alpha\partial^\alpha\pi\eta_{\mu\nu} - \partial_\mu\partial_\nu\pi) = (\Box h - \partial_\mu\partial_\nu h^{\mu\nu})\pi \tag{3.1}$$

Recall that the Ricci scalar in the weak field limit corresponds to $R = \partial_\mu\partial_\nu h^{\mu\nu} - \Box h$, such that the covariantization of (3.1) would give rise to

$$(\Box h - \partial_\mu\partial_\nu h^{\mu\nu})\pi \underbrace{\rightarrow}_{\text{cov.}} -R\pi \tag{3.2}$$

L. Heisenberg, *Theoretical and Observational Consistency of Massive Gravity*,
Springer Theses, DOI 10.1007/978-3-319-18935-2_3

After applying the same procedure to the other two interactions (even though more cumbersome) we obtain the following correspondences

$$h^{\mu\nu}X^{(1)}_{\mu\nu} \longleftrightarrow -\pi R \tag{3.3}$$

$$h^{\mu\nu}X^{(2)}_{\mu\nu} \longleftrightarrow -\partial_\mu\pi\partial_\nu\pi G^{\mu\nu} \tag{3.4}$$

$$h^{\mu\nu}X^{(3)}_{\mu\nu} \longleftrightarrow -\partial_\mu\pi\partial_\nu\pi\Pi_{\alpha\beta}L^{\mu\alpha\nu\beta}\,, \tag{3.5}$$

which relate the decoupling limit of massive gravity to some scalar-tensor interactions. The tensors $G_{\mu\nu}$ and $L^{\mu\alpha\nu\beta}$ are the Einstein and the dual Riemann tensors respectively

$$G^{\mu\nu} = R_{\mu\nu} - \frac{1}{2}Rg_{\mu\nu} \tag{3.6}$$

$$\begin{aligned} L^{\mu\alpha\nu\beta} &= 2R^{\mu\alpha\nu\beta} + 2(R^{\mu\beta}g^{\nu\alpha} + R^{\nu\alpha}g^{\mu\beta} - R^{\mu\nu}g^{\alpha\beta} - R^{\alpha\beta}g^{\mu\nu}) \\ &\quad + R(g^{\mu\nu}g^{\alpha\beta} - g^{\mu\beta}g^{\nu\alpha})\,. \end{aligned} \tag{3.7}$$

Thus, the covariantization of the decoupling limit Lagrangian (2.1) gives birth to the following proxy theory

$$\mathcal{L} = \sqrt{-g}\left(M_{Pl}^2 R + \mathcal{L}^\pi(\pi, g_{\mu\nu}) + \mathcal{L}^{matter}(\psi, g_{\mu\nu})\right)\,, \tag{3.8}$$

where the Lagrangian for π is

$$\mathcal{L}^\pi = M_{Pl}\left(-\pi R - \frac{a_2}{\Lambda^3}\partial_\mu\pi\partial_\nu\pi G^{\mu\nu} - \frac{a_3}{\Lambda^6}\partial_\mu\pi\partial_\nu\pi\Pi_{\alpha\beta}L^{\mu\alpha\nu\beta}\right)\,, \tag{3.9}$$

and $\mathcal{L}^{matter}$ is the Lagrangian for the matter fields ψ living on the geometry. This proxy theory represents a theory of GR on top of which a new scalar degree of freedom is added, which is non-minimally coupled to gravity. This theory can now be considered by its own and the modified equations of motion can be studied. Actually, this form of tensor structure has been first discussed by Horndeski (1974) in the context of the most general scalar-tensor theory. However, in difference to this work, the interesting point is that we obtained these interactions as a direct outcome of massive gravity. The variation of the action with respect to $g_{\mu\nu}$ yields the modified Einstein equation as

$$G_{\mu\nu} = M_{Pl}T^\pi_{\mu\nu} + T^{matter}_{\mu\nu} \tag{3.10}$$

with

$$T^\pi_{\mu\nu} = T^{\pi(1)}_{\mu\nu} - \frac{a_2}{\Lambda^3}T^{\pi(2)}_{\mu\nu} - \frac{a_3}{\Lambda^6}T^{\pi(3)}_{\mu\nu} \tag{3.11}$$

and

$$\begin{aligned} T^{\pi(1)}_{\mu\nu} &= X^{(1)}_{\mu\nu} + \pi G_{\mu\nu} \\ T^{\pi(2)}_{\mu\nu} &= X^{(2)}_{\mu\nu} + \frac{1}{2} L_{\mu\alpha\nu\beta}\partial^\alpha\pi\partial^\beta\pi + \frac{1}{2} G_{\mu\nu}(\partial\pi)^2 \\ T^{\pi(3)}_{\mu\nu} &= X^{(3)}_{\mu\nu} + \frac{3}{2} L_{\mu\alpha\nu\beta}\Pi^{\alpha\beta}(\partial\pi)^2 \end{aligned} \tag{3.12}$$

The structure of the Einstein and Riemann dual tensor ensures that π enters at most with two derivatives in the stress-energy tensor. Furthermore, the fact that G^{00}, G^{0i}, L^{0i0j} and L^{0ikj} have at most one time-derivative guarantees the propagation of constraints.

Since we are not in the Einstein frame, these stress-energy tensors are only transverse on-shell, and satisfy the relation, $D_\mu T^\mu_\nu = \partial_\nu \pi \mathcal{E}_\pi$ where $\mathcal{E}_\pi$ is the equation of motion with respect to π. Since both the Einstein tensor and the Riemann dual tensor are transverse, this equation of motion is also at most second order in derivative,

$$\begin{aligned} \mathcal{E}_\pi &= R + \frac{2a_2}{\Lambda^3} G^{\mu\nu}\Pi_{\mu\nu} \\ &+ \frac{3a_3}{\Lambda^6} L^{\mu\alpha\nu\beta}(\Pi_{\mu\nu}\Pi_{\alpha\beta} + R^\gamma{}_{\beta\alpha\nu}\partial_\gamma\pi\partial_\mu\pi) = 0 \end{aligned} \tag{3.13}$$

where we have used the fact that

$$D_\nu D_\alpha D_\beta \pi L^{\mu\alpha\nu\beta} = -R^\gamma{}_{\beta\alpha\nu}\partial_\gamma\pi L^{\mu\alpha\nu\beta} = \tfrac{1}{4}\partial^\mu\pi\mathcal{L}_{GB} \tag{3.14}$$

with the Gauss-Bonnet term $\mathcal{L}_{GB} = R^2 + R^2_{\mu\alpha\nu\beta} - 4R^2_{\mu\nu}$. In the following, we study the resulting cosmology in this proxy theory and address the question of whether or not self-accelerating solutions exist.

3.1 de Sitter Solutions

In what follows, we focus on the cosmology of the covariantized theory (3.8), (3.9), and focus for that on a FRW background with scale factor $a(t)$ and Hubble parameter H. The resulting effective energy density and pressure for the field π are then

$$\rho^\pi = M_{Pl}(6H\dot\pi + 6H^2\pi - \frac{9a_2}{\Lambda^3}H^2\dot\pi^2 - \frac{30a_3}{\Lambda^6}H^3\dot\pi^3) \tag{3.15}$$

$$\begin{aligned} P^\pi &= 2M_{Pl}\left[\frac{6a_3}{\Lambda^6}H\dot\pi^2(\dot\pi(\dot H + H^2) + \frac{3}{2}H\ddot\pi) + \frac{a_2}{2\Lambda^3}\dot\pi(\dot\pi(3H^2 + 2\dot H) + 4H\ddot\pi) \right. \\ &\left. - (\pi(3H^2 + 2\dot H) + 2H\dot\pi + \ddot\pi)\right], \end{aligned} \tag{3.16}$$

and the equation of motion for π (3.13) in the FRW space-time is equivalent to

$$\frac{6a_2}{\Lambda^3}\left(3H^3\dot{\pi}+2H\dot{H}\dot{\pi}+H^2\ddot{\pi}\right)+\frac{18a_3}{\Lambda^6}\left(3H^2\dot{H}\dot{\pi}^2+3H^4\dot{\pi}^2+2H^3\dot{\pi}\ddot{\pi}\right)=R. \tag{3.17}$$

This expression can be rewritten more compactly

$$\ddot{\phi}+3H\dot{\phi}-R=0 \tag{3.18}$$

after the field redefinition of the form

$$\dot{\phi}=H^2\left(\frac{6a_2}{\Lambda^3}\dot{\pi}+\frac{18a_3}{\Lambda^6}\dot{\pi}^2H\right). \tag{3.19}$$

3.1.1 Self-accelerating Solution

Now, we would like to study the self-acceleration solution with $H=\text{const}$ and $\dot{H}=0$ and $T_{\mu\nu}=0$ (i.e. the cosmological constant is set to zero). For the π field we make the ansatz $\dot{\pi}=q\frac{\Lambda^3}{H}$. Furthermore, we assume that we are in a regime where $H\pi\ll\dot{\pi}$ so that we can neglect terms proportional to π and consider only the terms including $\dot{\pi}$ or $\ddot{\pi}$. Thus, the Friedmann and field equations can be recast into

$$H^2=\frac{m^2}{3}(6q-9a_2q^2-30a_3q^3) \tag{3.20}$$

$$H^2(18a_2q+54a_3q^2-12)=0\,. \tag{3.21}$$

Assuming $H\neq 0$, the field equation then imposes,

$$q=\frac{-a_2\pm\sqrt{a_2^2+8a_3}}{6a_3} \tag{3.22}$$

while the Friedmann equation (3.20) sets the Hubble constant of the self-accelerated solution. Similar to what we had found in the decoupling limit, our proxy theory admits a self-accelerated solution, with the Hubble parameter set by the graviton mass. For the stability condition of this self-accelerating solution the first constraint we have is to demand $H>0$. The other constraint comes from the stability condition for perturbations on the background which we discuss in what follows.

3.1.2 Stability Conditions

Our proxy theory exhibits a self-accelerating solution with $H^2 \approx m^2$. The next step consists of studying the perturbations on this background and their stability constraints. For this purpose, we consider perturbations on top of the background solution in the following way

$$\pi = \pi_0(t) + \delta\pi(t, x, y, z) \tag{3.23}$$

The second order action for the perturbations is

$$\begin{aligned}\mathcal{L}_{\delta\pi} = &-\frac{a_2}{\Lambda^3}\partial_\mu\delta\pi\partial_\nu\delta\pi G^{\mu\nu} - \frac{a_3}{\Lambda^6}\partial_\mu\delta\pi\partial_\nu\delta\pi\Pi^{(0)}_{\alpha\beta}L^{\mu\alpha\nu\beta}\\ &-\frac{2a_3}{\Lambda^6}\partial_\mu\pi_0\partial_\nu\delta\pi D_\alpha D_\beta\delta\pi L^{\mu\alpha\nu\beta},\end{aligned} \tag{3.24}$$

which can be written in the form

$$\mathcal{L} = K_{tt}\left(\delta\dot{\pi}^2 - \frac{c_s^2}{a^2}(\nabla\delta\pi)^2\right) \tag{3.25}$$

where

$$K_{tt} = -\frac{3M_{Pl}a^3H^2}{\Lambda^3}\left(a_2 + \frac{6a_3H}{\Lambda^3}\dot{\pi}\right) \tag{3.26}$$

and

$$c_s^2 = \frac{1}{3}\left(2 + \frac{a_2\Lambda^3}{a_2\Lambda^3 + 6a_3H\dot{\pi}}\right). \tag{3.27}$$

The condition for the stability is then given by $K_{tt} > 0$, $c_s^2 > 0$ and $H^2 > 0$, which are fulfilled if

$$a_2 > 0 \quad \text{and} \quad 0 > a_3 > -\frac{1}{8}a_2^2. \tag{3.28}$$

To compare this result with the condition obtained in the decoupling limit in Eq. (2.13), we first mention that $a_2^{\text{Proxy}} = -2a_2^{\text{DL}}$ and $a_3^{\text{Proxy}} = a_3^{\text{DL}}$. In terms of the parameters used in the decoupling limit, we need to compare the conditions $a_2^{\text{Proxy}} < 0$, $a_3^{\text{Proxy}} > -\frac{1}{8}(a_2^{\text{Proxy}})^2$ to the conditions $a_2^{\text{DL}} < 0$ and $-\frac{2}{3}(a_2^{\text{DL}})^2 < a_3^{\text{DL}} < -\frac{1}{2}(a_2^{\text{DL}})^2$. We see that Proxy theory is less constraining but is still within the parameter space derived in the decoupling limit. It is not surprising that the stability condition in the decoupling limit and in the covariantized theory do not coincide totally as we have explicitly broken the symmetry when getting the proxy.

It is also worth pointing out that the self-accelerating solution by itself does not propagate any superluminal mode in the approximation we used, since $2/3 < c_s^2 < 1$.

We emphasize as well that the constant $\dot{\pi}$ solution is a dynamical attractor. For this we just consider time dependent perturbations $\pi(t) = \pi_0(t) + \delta\pi(t)$ which is a special case of (3.23) fulfilling the same stability conditions. The equation of motion for perturbations simplifies to[1]

$$\partial_t (a^3 \delta\dot{\pi}) = 0 \tag{3.29}$$

The solution for $\delta\dot{\pi}$ is given by

$$\delta\dot{\pi}(t) \sim a^{-3}. \tag{3.30}$$

Thus, these perturbations $\delta\pi(t)$ redshift away exponentially compared to the $\dot{\pi} =$ const self-accelerating solution. Therefore, the self-accelerating solution is an attractor.

3.1.3 *Degravitation*

More interestingly, one can wonder whether degravitation can be exhibited in these class of solutions. If one takes $\pi = \pi(t)$ and $H = 0$, it is straightforward to see that we obtain $\rho^{\pi} = 0$, so the field has absolutely no effect and cannot help the background to degravitate.

Interestingly, the interactions considered here are precisely of the same form as that studied in Charmousis et al. (2012). There as well, in the absence of spatial curvature $\kappa = 0$, the contribution from the scalar field vanishes if $H = 0$. Comparing with Charmousis et al. (2012), we can hence wonder whether the addition of spatial curvature $\kappa \neq 0$ in our proxy theory could help achieving degravitation, but relying strongly on spatial curvature brings concerns over instabilities which are beyond the scope of this thesis.

3.2 Cosmology

In the following we would like to discuss in more detail the interplay of all the constituents of the universe. We assume that matter, radiation and the scalar field π contribute to the total energy density of the universe.

$$H^2 = \frac{8\pi G}{3}(\rho^{\pi} + \rho^{rad} + \rho^{mat}) \tag{3.31}$$

[1]It is an attractor solution as long as $H\pi \ll \dot{\pi}$.

Consider the scalar field π as a perfect fluid with the effective energy density and pressure given by (3.15), (3.16). Thus, the equation of state parameter of this new field would be

$$\omega_\pi = \frac{1}{\mathcal{P}}\Big(-12a_3H^3\dot{\pi}^3 + 2\Lambda^3\dot{H}(2\Lambda^3\pi - a_2\dot{\pi}^2) + 3H^2(2\Lambda^6\pi - \dot{\pi}^2(a_2\Lambda^3 + 6a_3\ddot{\pi})) + 2\Lambda^6\ddot{\pi} + 4H\dot{\pi}(\Lambda^6 - 3a_3\dot{H}\dot{\pi}^2 - a_2\Lambda^3\ddot{\pi})\Big), \tag{3.32}$$

where $\mathcal{P} = 3H(2\Lambda^6\dot{\pi} - 10a_3H^2\dot{\pi}^3 + H(2\Lambda^6\pi - 3a_2\Lambda^3\dot{\pi}^2))$. At this point one should mention that the energy density for the π-field is not conserved but rather given by $D_\mu T^\mu_\nu = \partial_\nu\pi\mathcal{E}_\pi$ (where $\mathcal{E}_\pi$ is the equation of motion for π), which is not surprising since π is non-minimally coupled to gravity in the Jordan frame. Therefore, we can have $\omega_\pi < -1$.

In the following we will first assume that at early times in the evolution history of the Universe we can neglect the extra density coming from the helicity-0 ρ_π.[2] We will then check this assumption by plugging the solution for H back in the equation of motion for π. If we assume that at early times the radiation density dominates, we simply have

$$H^2 = \frac{8\pi G}{3}\rho_0^{rad}a^{-4} \qquad a \sim t^{1/2} \qquad \omega = 1/3 \tag{3.33}$$

During the radiation era, the dominant terms in the equation of motion for π are then $\frac{54a_3}{\Lambda^6}H^2\dot{H}\dot{\pi}^2 + \frac{54a_3}{\Lambda^6}H^4\dot{\pi}^2 + \frac{36a_3}{\Lambda^6}H^3\dot{\pi}\ddot{\pi} = 0$ which can be solved assuming the previous expression for H (3.33)

$$\pi_{rad} \sim t^{1.75} \qquad \text{yielding} \qquad \rho^\pi_{rad} \sim M_{Pl}t^{-1/4} \tag{3.34}$$

At later times when the matter dominated epoch starts we have

$$H^2 = \frac{8\pi G}{3}\rho_0^{mat}a^{-3} \qquad a \sim t^{2/3} \qquad \omega = 0 \tag{3.35}$$

Now the dominant terms in the equation of motion for π are $\frac{18a_2}{\Lambda^3}H^3\dot{\pi} + \frac{12a_2}{\Lambda^3}H\dot{H}\dot{\pi} + \frac{6a_2}{\Lambda^3}H^2\ddot{\pi} - 12H^2 - 6\dot{H} = 0$. We get for π this time

$$\pi^{mat} \sim c_2 \cdot t + \frac{t^2\Lambda^3}{4a_2} \tag{3.36}$$

[2] At early times the densities are large and the Vainshtein mechanism freezes out the helicity-0 mode.

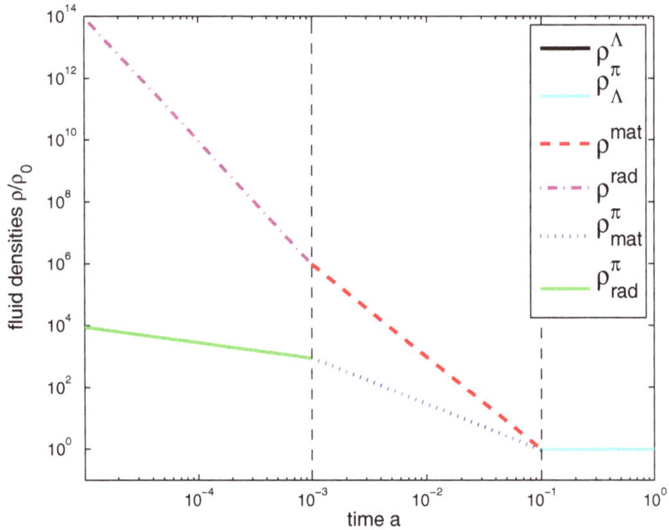

Fig. 3.1 Fluid densities $\rho^{rad} \sim a^{-4}$, $\rho^{mat} \sim a^{-3}$ and ρ_{π} during the epochs of radiation, matter and λ-domination normalised to today ρ_{π}. During the radiation domination the energy density for π goes as $\rho^{\pi}_{rad} \sim a^{-1/2}$ and during matter dominations as $\rho^{\pi}_{mat} \sim a^{-3/2}$ and is constant for later times $\rho^{\pi}_{\lambda} = \text{const}$

yielding

$$\rho^{\pi}_{mat} = c_2 M_{Pl} t^{-1} + \frac{3M_{Pl}(-14a_2^2 + 5a_3)\Lambda^3}{32a_2^3}. \tag{3.37}$$

Summarizing, during radiation domination the effective energy density for the π-field goes like $\rho^{\pi}_{rad} \sim t^{-1/4}$ while during matter domination as $\rho^{\pi}_{mat} \sim t^{-1}$ and approaches a constant at late time. As shown in Fig. 3.1, ρ^{π} can be neglected at early times where $\rho^{\pi}_{mat} \ll \rho^{mat}$ and $\rho^{\pi}_{rad} \ll \rho^{rad}$.

3.3 Structure Formation

We end the cosmological analysis by looking at the evolution of matter density perturbations. The density perturbations follow the evolution

$$\ddot{\delta}_m + 2H\dot{\delta}_m = \frac{\nabla^2\psi}{a^2} \tag{3.38}$$

where ψ is the Newtonian potential. The effects of π are all encoded in its contribution to the Poisson equation. Consider perturbations of π around its cosmological background solution $\pi(x,t) = \pi_0(t) + \phi(x,t)$. ϕ gives a contribution to the Newtonian potential of the form $\psi = \phi/M_{Pl}$. In the Newtonian approximation we have $|\dot{\phi}| \ll |\nabla\phi|$. The equation of motion for the scalar field in first order in ϕ is

$$-\frac{2a_2}{\Lambda^3}G^{\mu\nu}D_\mu D_\nu\phi - \frac{2a_3}{\Lambda^3}L_{\mu\alpha\nu\beta}(4\Pi_0^{\alpha\beta}D^\mu D^\nu\phi + 2R^{\gamma\beta\alpha\nu}\partial_\gamma\pi_0\partial^\mu\phi) = \delta R \quad (3.39)$$

which is equivalent to (neglecting $\dot{\phi}$)

$$\left[-\frac{2a_2}{\Lambda^3}(3H^2 + 2\dot{H}) + \frac{16a_3}{\Lambda^6}(2H^3\dot{\pi}_0 + 2\dot{H}H\dot{\pi}_0 + H\ddot{\pi}_0)\right]\frac{\nabla^2\phi}{a^2} = \delta R \quad (3.40)$$

and last but not least we need the trace of Einstein equation, (3.10)–(3.12). Perturbing the trace to first order, we get

$$-M_{Pl}\delta R = \left[3 - 2\frac{a_2}{\Lambda^3}(2H\dot{\pi}_0 + \ddot{\pi}_0) - \frac{3a_3}{\Lambda^6}(2\dot{H}\dot{\pi}_0^2 + 5H^2\dot{\pi}_0^2 + 4H\ddot{\pi}_0\dot{\pi}_0)\right]\frac{\nabla^2\phi}{a^2} + \frac{\delta T}{M_{Pl}} \quad (3.41)$$

To reach that point, we have neglected the perturbations of the curvature of the form $\delta R\pi_0$ as they are negligible compared to $M_{Pl}\delta R$ since we work in the regime where $\pi_0 \ll M_{Pl}$. As a first approximation, such terms are hence ignored.

The perturbations for the source is just given by $\delta T = -\rho_m\delta_m$ for non-relativistic sources, thus we have

$$\frac{\nabla^2\phi}{a^2} = \frac{\rho_m\delta_m}{3M_{Pl}\mathcal{Q}} \quad (3.42)$$

where $\mathcal{Q}$ stands for

$$\mathcal{Q} \equiv 1 - \frac{2a_2}{\Lambda^3}(2H\dot{\pi}_0 + \ddot{\pi}_0 + M_{Pl}(2\dot{H} + 3H^2)) - \frac{a_3}{\Lambda^6}\left(5H^2\dot{\pi}_0^2 + 2\dot{H}\dot{\pi}_0^2 + 4H\ddot{\pi}_0\dot{\pi}_0 - \frac{16M_{Pl}}{3}(2H^3\dot{\pi}_0 + 2H\dot{H}\dot{\pi}_0 + H^2\ddot{\pi}_0)\right). \quad (3.43)$$

Finally, the modified evolution equation for density perturbations is

$$\ddot{\delta}_m + 2H\dot{\delta}_m = \frac{\rho_m\delta_m}{M_{Pl}^2}\left(1 + \frac{1}{3\mathcal{Q}}\right). \quad (3.44)$$

Knowing the background configuration it is then relatively straightforward to derive the effect on structure formation. We recover the usual result that when the field is screened $H\dot{\pi} \gtrsim \Lambda^3$, i.e. Q is large, the extra force coming from the helicity-0 is negligible and the formation of structure is similar as in ΛCDM.

3.4 Covariantization from the Einstein Frame

So far we have studied our proxy theory in the Jordan frame. The unavoidable question arises whether our results remain the same in a different frame. Instead of covariantizing our Lagrangian in the Jordan frame, it is on an equal footing to go to the Einstein frame first where the Ricci scalar is not multiplied by the scalar field π and covariantize the theory at that stage. Our starting Lagrangian was

$$\mathcal{L}_{DL} = -\frac{1}{2}h^{\mu\nu}\mathcal{E}^{\alpha\beta}_{\mu\nu}h_{\alpha\beta} + h^{\mu\nu}\sum_{n=1}^{3}\frac{a_n}{\Lambda_3^{3(n-1)}}X^{(n)}_{\mu\nu}(\Pi) + \frac{1}{2M_{Pl}}h^{\mu\nu}T_{\mu\nu} \tag{3.45}$$

Remember, the mixing between the Ricci scalar and the scalar field was due to the covariantization of the mixed term $h^{\mu\nu}X^{(1)}_{\mu\nu}$, which we can diagonalize by performing the following change of variables

$$h_{\mu\nu} = \bar{h}_{\mu\nu} + \pi\eta_{\mu\nu} \,. \tag{3.46}$$

The Lagrangian then becomes

$$\begin{aligned}\mathcal{L}_{\rm diag} = & -\frac{1}{2}\bar{h}^{\mu\nu}\mathcal{E}^{\alpha\beta}_{\mu\nu}\bar{h}_{\alpha\beta} + \frac{3}{2}\pi\Box\pi + \frac{a_2}{\Lambda^3}\bar{h}^{\mu\nu}X^{(2)}_{\mu\nu} - \frac{3}{2}\frac{a_2}{\Lambda^3}\Box\pi(\partial\pi)^2 \\ & + \frac{a_3}{\Lambda^6}\bar{h}^{\mu\nu}X^{(3)}_{\mu\nu} - 2\frac{a_3}{\Lambda^6}(\partial\pi)^2([\Pi^2] - (\Box\pi)^2) + \frac{1}{2M_{Pl}}(\bar{h}_{\mu\nu} + \pi\eta_{\mu\nu})T^{\mu\nu}\end{aligned} \tag{3.47}$$

Covariantizing this action is straightforward. We use again the correspondences in (3.3) and it has been shown explicitly that the covariantization of the Galileon interactions

$$\begin{aligned}\Box\pi(\partial\pi)^2 & \underbrace{\rightarrow}_{\text{cov.}} \Box\pi(\partial\pi)^2 \\ -2(\partial\pi)^2([\Pi^2] - \Box\pi^2) & \underbrace{\rightarrow}_{\text{cov.}} 2(\partial\pi)^2\left((\Box\pi)^2 - [\Pi^2] - \frac{1}{4}(\partial\pi)^2R\right)\end{aligned} \tag{3.48}$$

does not yield any ghostlike instabilities (Deffayet et al. 2009; de Rham and Tolley 2010). Thus, the covariantized action in the Einstein frame is simply given by

$$\mathcal{L}_{\rm Einst} = M_{\rm Pl}^2 R + \frac{3}{2}\pi\Box\pi - \frac{a_2 M_{\rm Pl}}{\Lambda^3}\partial_\mu\pi\partial_\nu\pi G^{\mu\nu}$$
$$- \frac{3}{2}\frac{a_2}{\Lambda^3}\Box\pi(\partial\pi)^2 + 2\frac{a_3}{\Lambda^6}(\partial\pi)^2\left((\Box\pi)^2 - [\Pi^2] - \frac{1}{4}(\partial\pi)^2 R\right)$$
$$- \frac{a_3 M_{\rm Pl}}{\Lambda^6}\partial_\mu\pi\partial_\nu\pi\Pi_{\alpha\beta}L^{\mu\alpha\nu\beta} + \mathcal{L}_m(\psi, (1+\pi)g_{\mu\nu}) \qquad (3.49)$$

As you can see, there still remain non-minimal couplings. In these types of theories one can never fully go to the Einstein frame. Similarly as before, the properties of $G^{\mu\nu}$ and $L^{\mu\alpha\nu\beta}$ ensure that their equations of motion lead at most to second order derivative terms. To find a self-accelerating solution we set again a pure de Sitter metric, with $\dot{\pi} = q\Lambda^3/H$. The Friedmann and the field equations then take the form

$$M_{\rm Pl}^2 H^2 = 3M_{\rm Pl}a_2\Lambda^3 q^2 + 10M_{\rm Pl}a_3\Lambda^3 q^3 - \frac{1}{2}\frac{\Lambda^6}{H^2}q^2$$
$$+ 3a_2\frac{\Lambda^6}{H^2}q^3 + 15a_3\frac{\Lambda^6}{H^2}q^4 + \Lambda^3(-1 + 3q(a_2 + 4a_3 q))$$
$$+ 2M_{\rm Pl}(a_2 + 3a_3 q)H^2 = 0 \qquad (3.50)$$

When comparing the above Friedmann and the field equations with the one we had in the Jordan frame (3.20), we see significant differences coming from the extra terms which were not there in the Jordan frame. These terms yield Friedmann and field equations proportional to q^4 and H^4 which are more difficult to solve.

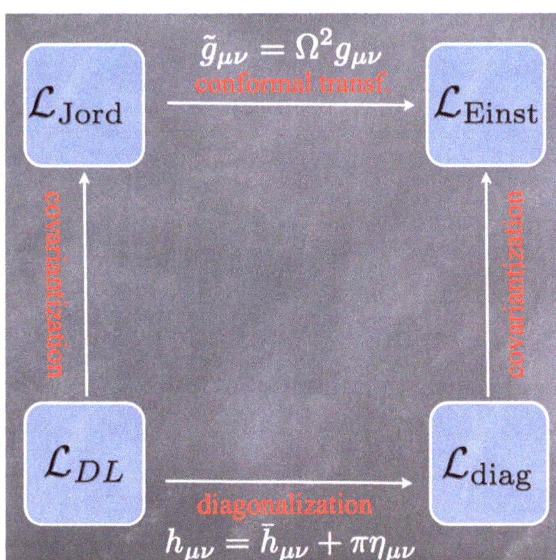

Fig. 3.2 Covariantization from different frames

For fairness, we should compare both actions in the same frame (Fig. 3.2). We do so by performing a conformal transformation on the action (3.9):

$$\tilde{g}_{\mu\nu} = \Omega^2 g_{\mu\nu} \qquad \text{with} \qquad \Omega^2 = \left(1 - \frac{\pi}{M_{Pl}}\right). \tag{3.51}$$

Using the fact that

$$\begin{aligned}
\Omega_{;\mu} &= -\frac{1}{M_{Pl}\Omega}\partial_\mu \pi \\
\Omega_{;\mu\nu} &= -\frac{1}{M_{Pl}^2\Omega^3}\partial_\mu \pi \partial_\nu \pi - \frac{1}{M_{Pl}\Omega}\Pi_{\mu\nu} \\
(\ln \Omega)_{;\mu\nu} &= -\Omega^{-2}\Omega_{;\mu}\Omega_{;\nu} + \Omega^{-1}\Omega_{;\mu\nu}
\end{aligned} \tag{3.52}$$

we can write down how the following important quantities transform

$$\begin{aligned}
\sqrt{-g} &= \Omega^{-4}\sqrt{-\tilde{g}} \\
R &= \Omega^2\tilde{R} - \frac{18\Omega^{-2}}{M_{Pl}^2}(\tilde{\partial}\pi)^2 - \frac{6}{M_{Pl}}\tilde{\Box}\pi \\
R_{\mu\nu} &= \tilde{R}_{\mu\nu} - \frac{2\Omega^{-4}}{M_{Pl}^2}\tilde{\partial}_\mu \pi \tilde{\partial}_\nu \pi - \frac{2\Omega^{-2}}{M_{Pl}}\tilde{\nabla}_\mu \tilde{\partial}_\nu \pi \\
&\quad - \frac{4\Omega^{-4}}{M_{Pl}^2}\tilde{g}_{\mu\nu}(\tilde{\partial}\pi)^2 - \frac{\Omega^{-2}}{M_{Pl}}\tilde{g}_{\mu\nu}\tilde{\Box}\pi \\
G_{\mu\nu} &= \tilde{G}_{\mu\nu} - \frac{2\Omega^{-4}}{M_{Pl}^2}\tilde{\partial}_\mu \pi \tilde{\partial}_\nu \pi - \frac{2\Omega^{-2}}{M_{Pl}}\tilde{\nabla}_\mu \tilde{\partial}_\nu \pi \\
&\quad + \frac{5\Omega^{-4}}{M_{Pl}^2}\tilde{g}_{\mu\nu}(\tilde{\partial}\pi)^2 + 2\frac{\Omega^{-2}}{M_{Pl}}\tilde{g}_{\mu\nu}\tilde{\Box}\pi
\end{aligned} \tag{3.53}$$

For simplicity we consider the case for which $a_3 = 0$, so under this conformal transformation the covariantized action (3.9) becomes

$$\begin{aligned}
\mathcal{L}_{Jord} &= M_{Pl}^2\tilde{R} - \frac{3}{2}\Omega^{-4}(\tilde{\partial}\pi)^2 - \frac{a_2 M_{Pl}}{\Lambda^3}\left(\tilde{\partial}^\mu \pi \tilde{\partial}^\nu \pi \tilde{G}_{\mu\nu}\right) \\
&\quad - \frac{a_2 M_{Pl}}{\Lambda^3}\left(\frac{3}{2}\frac{\Omega^{-2}}{M_{Pl}}(\tilde{\partial}\pi)^2\tilde{\Box}\pi + \frac{5}{4}\frac{\Omega^{-4}}{M_{Pl}^2}(\tilde{\partial}\pi)^4\right)
\end{aligned} \tag{3.54}$$

In the limit where $\pi \ll M_{Pl}$ we have then finally the following expression

$$\mathcal{L}_J = M_{Pl}^2 R + \frac{3}{2}\pi\Box\pi - \frac{a_2 M_{Pl}}{\Lambda^3}\partial_\mu\pi\partial_\nu\pi G^{\mu\nu} - \frac{3}{2}\frac{a_2}{\Lambda^3}\Box\pi(\partial\pi)^2\,, \qquad (3.55)$$

which coincides with the theory obtained from the Einstein frame, (3.49) with $a_3 = 0$. So within the regime of validity of our results, our conclusions are independent of the choice of frame. However beyond the regime $\pi \ll M_{Pl}$ the theory originally constructed from the Jordan frame could violate the null energy condition from the term proportional to $(\partial\pi)^4$.

3.5 Proxy Theory as a Subclass of Horndeski Scalar-Tensor Theories

As we mentioned before, the proxy theory is a subclass of Horndeski scalar-tensor theories which describe the most general scalar tensor interactions with second order equations of motion. In the following we will relate the general functions of he Horndeski interactions with the Proxy theory. The Horndeski theory is given by the action

$$s = \int d^4x\sqrt{-g}\left(\sum_{i=2}^{5}\mathcal{L}_i + \mathcal{L}_m\right) \qquad (3.56)$$

with

$$\begin{aligned}
\mathcal{L}_2 &= K(\pi, X)\\
\mathcal{L}_3 &= -G_3(\pi, X)[\Pi]\\
\mathcal{L}_4 &= G_4(\pi, X)R + G_{4,X}\left([\Pi]^2 - [\Pi^2]\right)\\
\mathcal{L}_5 &= G_5(\pi, X)G_{\mu\nu}\Pi^{\mu\nu} - \frac{1}{6}G_{5,X}\left([\Pi]^3 - 3[\Pi][\Pi^2] + 2[\Pi^3]\right) \qquad (3.57)
\end{aligned}$$

where the arbitrary functions K, G_3, G_4 and G_5 depend on the scalar field π and its derivatives $X = -\frac{1}{2}(\partial\pi)^2$ and furthermore $G_{i,X} = \partial G_i/\partial X$ and $G_{i,\pi} = \partial G_i/\partial\pi$. Our proxy theory corresponds to the case for which the above functions take the concrete following forms

$$\begin{aligned}
K(\pi, X) &= 0\\
G_3(\pi, X) &= 0\\
G_4(\pi, X) &= M_{Pl}^2 - M_{Pl}\pi - \frac{M_{Pl}}{\Lambda^3}a_2 X\\
G_5(\pi, X) &= 3\frac{M_{Pl}}{\Lambda^6}a_3 X \qquad (3.58)
\end{aligned}$$

In the previous section we have studied the stability of the scalar field perturbation on top of the self-accelerating background and ignored the tensor perturbations so far. We can study the tensor perturbations as well and put constraints on the two free parameters. For that we can use the general expression for the tensor perturbations in the Horndeski theory studied already in Kimura and Yamamoto (2012)

$$\mathcal{L} = K_{tt}\left(\dot{h}_{ij}^2 - \frac{c_s^2}{a^2}(\nabla h_{ij})^2\right) \tag{3.59}$$

with

$$\begin{aligned} K_{tt} &= G_4 - 2XG_{4,X} - X\left(H\dot{\pi}G_{5,X} - G_{5,\pi}\right) \\ c_s^2 &= \frac{G_4 - X\left(\ddot{\pi}G_{5,X} + G_{5,\pi}\right)}{K_{tt}} \end{aligned} \tag{3.60}$$

with the functions K, G_3, G_4 and G_5 given as in (3.58). The sound speed of the tensor perturbations depends explicitly on the background solution for π. Similarly as in the previous sections we can put constraints on the two parameters of the theory by demanding the ghost absence $K_{tt} > 0$ and Laplacian stability $c_s^2 > 0$ with the self accelerating background solution $\dot{\pi} = q\Lambda^3/H$ in the regime $H\pi \ll \dot{\pi}$. This gives as constraints

$$a_2 > 0 \qquad \text{and} \qquad a_3 > a_2^2. \tag{3.61}$$

These constraints are less constraining than the constraints coming from the stability analysis of the scalar perturbations of π but are still within the same parameter space. On top of these stability constraints we can tighten up more the parameters using the constraints from gravitational Cherenkov radiation which is

$$c_s^2 > 1 - 2 \times 10^{-15} \tag{3.62}$$

This constraint tighten up further the parameters space to be in the regime $a_2 > 0$ and $a_3 < (3 + 5 \times 10^{-15})a_2^2$.

3.6 Critical Assessment of the Proxy Theory

We would like now undertake a critical assessment of the analysis performed so far in this chapter, which were the outcomes in de Rham and Heisenberg (2011). The following results obtained in this section are unpublished results and serve as clarification of the validity of our previous results. The first limitation is the fact that the solutions found in Sect. 3.1.1 are only valid in the approximation $H\pi \ll \dot{\pi}$ we used. There, we have argued that de Sitter is a legitimate solution when such an approximation holds. In the following we will study the validity of this approximation in more detail. In a pure de Sitter background with constant expansion rate H_{dS}, the exact homogeneous field equation reads

$$\frac{6H_{dS}^2}{\Lambda^3}\left(a_2 + 6a_3\frac{H_{dS}}{\Lambda^3}\dot{\pi}\right)\ddot{\pi} + 18\frac{H_{dS}^3}{\Lambda^3}\left(a_2 + 3a_3\frac{H_{dS}}{\Lambda^3}\dot{\pi}\right)\dot{\pi} = 12H_{dS}^2. \tag{3.63}$$

In our previous analysis, we solved this equation together with Friedman equation by using the approximation $\pi H \ll \dot{\pi}$ and the ansatz of constant $\dot{\pi}$. However, this equation can actually be exactly solved and the corresponding solution exhibits the two following branches for $\dot{\pi}$:

$$\dot{\pi} = \frac{-a_2\Lambda^3 \pm e^{-\frac{3}{2}H_{dS}t}\sqrt{4a_3e^{C_1} + (a_2^2 + 8a_3)e^{3H_{dS}t}\Lambda^6}}{6a_3H_{dS}} \tag{3.64}$$

with C_1 an integration constant. At late times, one can easily see that $\dot{\pi}$ evolves towards the constant value

$$\dot{\pi}(t \gg H_{dS}^{-1}) \simeq -\frac{\Lambda^3}{6a_3H_{dS}}\left[a_2 \pm \sqrt{a_2^2 + 8a_3}\right]. \tag{3.65}$$

This coincides with our previous finding when assuming the ansatz $\dot{\pi} = q\frac{\Lambda^3}{H_{dS}}$, showing that such a solution is indeed the attractor solution in a de Sitter background. It is important to notice that this solution has been obtained by assuming that the de Sitter background is not driven by the π field, but by some other independent effective cosmological constant. Now we want to study if such an effective cosmological constant can be generated by the π field itself so that de Sitter is an actual solution of the system. From the above solution for $\dot{\pi}$, it is straightforward to obtain the solution for π by means of a simple integration

$$\pi(t \gg H_{dS}^{-1}) \simeq -\frac{\Lambda^3}{6a_3H_{dS}}\left[a_2 \pm \sqrt{a_2^2 + 8a_3}\right]t + C_2 \tag{3.66}$$

where C_2 is another integration constant. If we plug this solution into the energy density of π (which gives the r.h.s. of Friedman equation), we obtain

$$\rho_\pi \simeq \frac{M_p\Lambda^3}{18}\left[108C_2\frac{H_{dS}^2}{\Lambda^3} + \left(\frac{a_2^3}{a_3^2} + 6\frac{a_2}{a_3}\right) \pm \left(\frac{a_2^2 + 2a_3}{a_3^2}\right)\sqrt{a_2^2 + 8a_3}\right] - \frac{M_p\Lambda^3}{a_3}\left(a_2 \pm \sqrt{a_2^2 + 8a_3}\right)H_{dS}t. \tag{3.67}$$

At early times when $H_{dS}t \ll 1$ we can neglect the second term in this expression, the energy density of the π field is approximately constant, as it corresponds to a de

Sitter solution. However, we must keep in mind that this solution is actually valid at late times and, in that case, the second term growing linearly with time drives the energy density evolution and, thus, de Sitter cannot be the solution. This also agrees with the fact that the condition $\pi H \ll \dot{\pi}$ will be eventually violated at late times because the scalar field grows in time, whereas H and $\dot{\pi}$ are assumed to be constant. One might think that a way out would be to tune the parameters so that $a_2 \pm \sqrt{a_2^2 + 8a_3} = 0$. However, the only solution to this equation is $a_3 = 0$, which represents a singular value. In fact, if we take the limit $a_3 \to 0$ in the above solution, we obtain $\rho_\pi \to 6C_2 H_{dS} M_p + 4M_p \Lambda^3 H_{dS} t/a_2$ so the growing term remains. From this simple analysis, it seems that de Sitter cannot exist as an attractor solution of the phase map, but it can only represent transient regimes. This can in turn be useful for inflationary models where the accelerated expansion needs to end, but it is less appealing as dark energy model.

In the following, we will make this simple analysis more rigorous and look at it in more detail. In order to obtain a general overview of the class of cosmological solutions that one can expect to find in the proxy theory, we shall perform a dynamical system analysis. This will give us the critical points of the cosmological equations as well as their stability. The first step to perform the dynamical system analysis will be to obtain the equations to be analysed. Since we are interested in cosmological solutions, the metric will be assumed to take the FRW form with flat spatial sections. The most convenient time variable for the analysis will be the number of e-folds $N \equiv \ln a$. The equation of motion for the π field in terms of this time variable is given by

$$\left(a_2 + 6a_3 H^2 \frac{\pi'}{\Lambda^3}\right)\pi'' + 3\left[a_2\left(1 + \frac{H'}{H}\right) + \frac{a_3 H^2}{\Lambda^3}\left(3 + 5\frac{H'}{H}\right)\pi'\right]\pi'$$
$$= 2\frac{\Lambda^3}{H^2}\left(1 + \frac{H'}{2H}\right) \tag{3.68}$$

where the prime denotes derivative with respect to N. In addition to this equation, we also need the corresponding Einstein equations, which in our case are given by

$$H^2 = \frac{1}{6M_p^2}\rho_\pi \tag{3.69}$$

$$2HH' + 3H^2 = -\frac{1}{2M_p^2}p_\pi \tag{3.70}$$

where we have used that $dN = Hdt$ and ρ_π and p_π are the energy density and pressure of the π field expressed in terms of N. We have now 3 equations for the two variables π and H. Of course, not all three equations are independent. In order to reduce these equations to the form of an autonomous system, we will first use the Friedman constraint to obtain an expression for π in terms of π' and H. The resulting expression will constitute a constraint for π and will allow us to get rid of its dependence in the remaining equations so that we end up with dependence only on H, H′, π′ and π″. This will result very useful since it reduces the number of variables in our autonomous system. In fact, we can use $y \equiv \pi'$ as one of our dynamical variables and, then, we have a system of two first order differential equations for y and H. After some simple algebra, one can reduce the equations to the following autonomous system:

$$\begin{aligned} \frac{dy}{dN} &= -\frac{1+3b_2H^2y+(25b_3-9b_2^2)H^4y^2-87b_2b_3H^6y^3-180b_3^2H^8y^4}{1-6b_2H^2y+6(b_2^2-5b_3)H^4y^2+52b_2b_3H^6y^3+105b_3^2H^8y^4}y \\ \frac{dH}{dN} &= -\frac{2-8b_2H^2y+(9b_2^2-33b_3)H^4y^2+72b_2b_3H^6y^3+135b_3^2H^8y^4}{1-6b_2H^2y+6(b_2^2-5b_3)H^4y^2+52b_2b_3H^6y^3+105b_3^2H^8y^4}H \end{aligned} \tag{3.71}$$

where we have introduced the rescaled parameters $b_2 \equiv \frac{M_p^3}{\Lambda^3}a_2$ and $b_3 \equiv \frac{M_p^6}{\Lambda^6}a_3$. One can immediately see that $H = y = 0$ is a stable critical point which is independent of the parameters and corresponds to the vacuum Minkowski solution. For the remaining critical points, we need to solve the equations

$$\begin{aligned} 1+3b_2H^2y+(25b_3-9b_2^2)H^4y^2-87b_2b_3H^6y^3-180b_3^2H^8y^4 &= 0, \\ 2-8b_2H^2y+(9b_2^2-33b_3)H^4y^2+72b_2b_3H^6y^3+135b_3^2H^8y^4 &= 0. \end{aligned}$$

To solve these equations, it will be convenient to introduce a new rescaling as $\hat{y} \equiv H^2yb_2$ and the new constant $c_3 \equiv b_3/b_2^2 = a_3/a_2^2$. Then, the previous equations can be written in the simpler form

$$1+3\hat{y}+(25c_3-9)\hat{y}^2-87c_3\hat{y}^3-180c_3^2\hat{y}^4 = 0, \tag{3.72}$$

$$2-8\hat{y}+(9-33c_3)\hat{y}^2+72c_3\hat{y}^3+135c_3^2\hat{y}^4 = 0. \tag{3.73}$$

As we can see, we have an overdetermined system of equations so that solutions cannot be found for arbitrary c_3. In fact, the above equations can be solved for $\hat{y}$ and c_3 in order to obtain the models with additional critical points. Remarkably, there is only one real solution for these equations and is given by $c_3 \simeq 0.094$ and $\hat{y} \simeq -3.99$. Notice that this in fact does not represent one single critical point for the autnomous system, but a curve of critical points in the plane (y, H). The obtained result implies that pure de Sitter does not correspond to a critical point of the proxy theory and can only exist as a transient regime, as we had anticipated from our previous simple analysis.

Another interesting feature of the autonomous system is the existence of separatrices in the phase map determined by the curve along which the denominators in (3.71) vanish, i.e.

$$1-6b_2H^2y+6(b_2^2-5b_3)H^4y^2+52b_2b_3H^6y^3+105b_3^2H^8y^4 = 0. \tag{3.74}$$

This curve can be simplified if we use our previously defined rescaled variable $\hat{y}$ and parameter c_3, in terms of which the separatrix is determined by

$$1 - 6\hat{y} + 6(1 - 5c_3)\hat{y}^2 + 52c_3\hat{y}^3 + 105c_3^2\hat{y}^4 = 0 \tag{3.75}$$

which is a quartic polynomial equation. Being the independent term and the highest power coefficient both positive, this equation does not always have real solutions so the separatrix does not exist for arbitrary parameters. Indeed, the previous equation determines a curve in the plane $(\hat{y}, c_3)$, which can be regarded as the function

$$c_3 = \frac{15 - 26\hat{y} \pm \sqrt{2}\sqrt{60 - 75\hat{y} + 23\hat{y}^2}}{105\hat{y}^2}. \tag{3.76}$$

This function has been plotted in Fig. 3.3. As we can see in that figure, the value of c_3 determines the number of real solutions and, therefore, the number of separatrices in the phase map of the autonomous system. We find that for $c_3 > 0$, the system always exhibits 4 separatrices. When $c_3 = 0$, the cubic and quartic terms of the separatrix equation vanish, so we only have two real solutions. In the cases with $0 > c_3 > -0.093$, the system has 4 separatrices again. When $-0.093 > c_3 > -0.215$, there are only 2 separatrices and, finally, for $c_3 < -0.215$, the equation has no real solutions and, therefore, it does not generate any separatrix. Special cases are $c_3 = -0.093$ with 3 separatrices and $c_3 = -0.215$ with only one separatrix. All this can be clearly seen in Fig. 3.3.

If the solutions of Eq. (3.76) are denoted by $\hat{y} = y_i^*$, then, the separatrices are given by the curves $y = b_2 y_i^*/H^2$ or, equivalently, $H = \pm\sqrt{b_2 y_i^*/y}$ in the phase map. Notice that, depending on the sign of $b_2 y_i^*$, the corresponding separatrix will only exist in the semi-plane $y > 0$ or $y < 0$ for $b_2 y_i^* > 0$ or $b_2 y_i^* < 0$ respectively. This can be seen in the examples shown in Fig. 3.4 where we have plotted the phase maps corresponding to two characteristic cases, namely, one with $c_3 = 1.5$ (which has 4

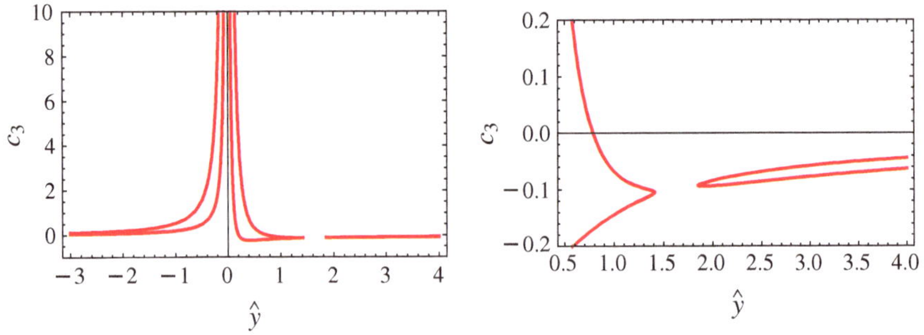

Fig. 3.3 In this plot we show the *curve* determined by Eq. (3.75) in the plane $(\hat{y}, c_3)$. The *left panel* shows a detail to see more clearly the structure of the corresponding area. As explained in the main text, the value of the parameter c_3 determines the number of separatrices in the phase map

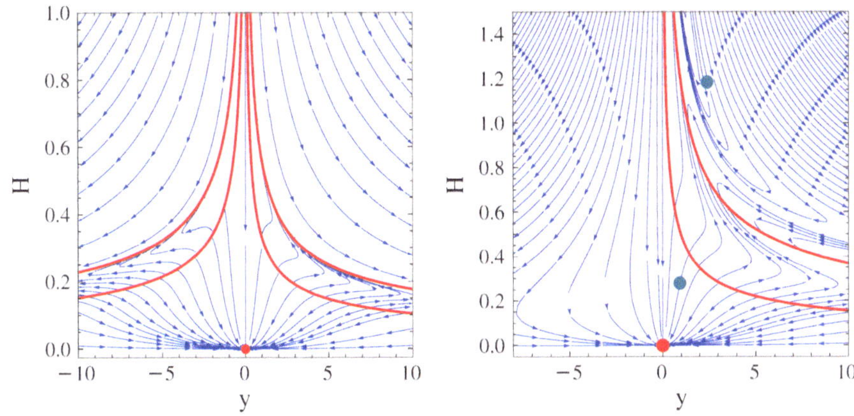

Fig. 3.4 In this figure we show two examples of phase map portraits of the dynamical autonomous system for $b_2 = 1$ and $b_3 = 1.5$ (with $c_3 = 1.5$) in the *left panel* and $b_2 = 1$ and $b_3 = -0.1$ (with $c_3 = -0.1$) in the *right panel*. These values have been chosen to show examples with $c_3 > 0$ (always with 4 separatrices) and $c_3 < 0$ with 2 separatrices (see main text and Fig. 3.3). The *red lines* represent the corresponding separatrices and the *red point* denotes the Minkowski vacuum solution. We can see that this solution is indeed an attractor. Concerning the attracting behaviour of the separatrices, we can see that the upper ones behave as attractors, whereas the lower ones act as repelers. In the *right panel*, we additionally indicate with *green points* the analytical solutions found in previous sections under the approximation $\pi H \ll \dot{\pi}$

separatrices and positive c_3) and one with $c_3 = -0.1$ (which has only 2 separatrices and negative c_3). One interesting feature that we can observe in both cases is the attracting nature of the upper separatrices, whereas the lower ones behave as repellers. Remarkably, the attracting separatrices do not behave as asymptotic attractors, but the trajectories actually hit the separatrix and the universe encounters a singularity.

The phase map shown in the right panel corresponds to parameters satisfying all the existence and stability requirements obtained in previous sections from the approximate analytical solutions. The green points in the phase map denote the solutions that we had identified with stable self-accelerating solutions. However, we can see now that the eventual attractor solution is not actually de Sitter but the Minkowksi vacuum solution. The stability condition for such a solution actually corresponds to the convergence of the nearby trajectories.

It is worthwhile pointing out once more that, although (quasi) de Sitter solutions do not exist as critical points in the phase maps, it is possible to have transient regimes with quasi de Sitter expansion. One possibility where such transient regimes can be found correspond to the trajectories above the upper separatrix in the right panel of Fig. 3.4. These trajectories initially evolve towards large values of y, but, at some point, there is a turnover where it goes towards smaller values of y. While this turnover is taking place, the value of H can remain nearly constant for some time and, thus, we can have a period of quasi de Sitter expansion. The number of e-folds corresponding to this transient regime depends on the parameters and the initial

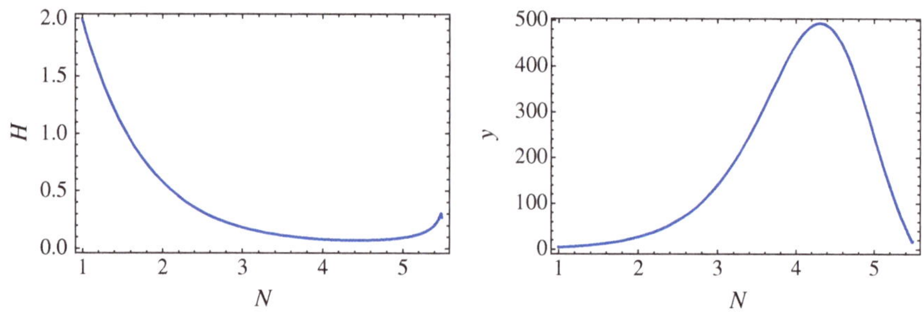

Fig. 3.5 In this figure we show the numerical solution for H (*left panel*) and y (*right panel*) with the initial conditions $H_{ini} = 2$ and $y_{ini} = 5$. We can see the transient period of quasi de Sitter expansion in the evolution of H corresponding to the turnover explained in the main text and how it lasts for barely 1−2 e-folds. In addition, we can see the discussed singularity corresponding to the moment when the trajectory reaches the separatrix at a finite number of e-folds

conditions, but it is generally quite small (see Fig. 3.5 where we plot the evolution of one particular solution).

In order to study the properties of the dynamical system near the separatrix, we will rewrite the autonomous system in terms of the variable $\hat{y}$, since, as suggested from our previous analysis, the equations will look simpler. In particular, the separatrices will become straight vertical lines in this variable and the behaviour of the trajectories near them can be straightforwardly studied. In such variables, the autonomous system reads

$$\begin{aligned} \frac{d\hat{y}}{dN} &= -\frac{5 - 13\hat{y} + (9 - 41c_3)\hat{y} + 57c_3\hat{y}^3 + 90c_3^2\hat{y}^4}{1 - 6\hat{y} + 6(1 - 5c_3)\hat{y}^2 + 52c_3\hat{y}^3 + 105c_3^2\hat{y}^4}\hat{y} \\ \frac{dH}{dN} &= -\frac{2 - 8\hat{y} + (9 - 33c_3)\hat{y}^2 + 72c_3\hat{y}^3 + 135c_3^2\hat{y}^4}{1 - 6\hat{y} + 6(1 - 5c_3)\hat{y}^2 + 52c_3\hat{y}^3 + 105c_3^2\hat{y}^4}H. \end{aligned} \tag{3.77}$$

As we anticipated, the equations look simpler in these variables. In particular, the equation for $\hat{y}$ completely decouples from the equation for the Hubble expansion rate. Near the separatrix located at y_s we can expand $\hat{y} = \hat{y}_s + \delta\hat{y}$ and obtain the leading terms of the above equations, given by

$$\frac{d\delta\hat{y}}{dN} = \frac{k_y}{\delta\hat{y}}, \qquad \frac{dH}{dN} = \frac{k_H}{\delta\hat{y}}H \tag{3.78}$$

with

$$k_y \equiv -\frac{5 - 13\hat{y}_s + (9 - 41c_3)\hat{y}_s + 57c_3\hat{y}_s^3 + 90c_3^2\hat{y}_s^4}{1 - 6 + 12(1 - 5c_3)\hat{y}_s + 156c_3\hat{y}_s^2 + 420c_3^2\hat{y}_s}\hat{y}_s \tag{3.79}$$

$$k_H \equiv -\frac{2 - 8\hat{y}_s + (9 - 33c_3)\hat{y}_s^2 + 72c_3\hat{y}_s^3 + 135c_3^2\hat{y}_s^4}{1 - 6 + 12(1 - 5c_3)\hat{y}_s + 156c_3\hat{y}_s^2 + 420c_3^2\hat{y}_s}. \tag{3.80}$$

Now, it is straightforward to read the conditions for the separatrix to attract the trajectories. Notice that the attracting or repelling nature of the separatrix will be the same from both sides of it. Thus, whenever k_y is negative, the separatrix will represent an attractor of the phase map, whereas it will be a repeller for positive k_y.

The equation for $\delta\hat{y}$ near the separatrix can be easily integrated to give

$$\delta\hat{y}(N) \simeq \pm\sqrt{2k_y N + C_y} \tag{3.81}$$

with C_y an integration constant and the two branches correspond to both sides of the speratrix. If the separatrix is an attractor, we have that k_y is negative and, therefore, the solution only exists until $N_s = -\frac{C_y}{2k_y}$, confirming our previous statement that the trajectories do not approach asymptotically the separatrix, but they hit it and end there. On the other hand, with the solution for $\delta\hat{y}$, we can also obtain the solution for H, which is given by

$$H(N) = C_H e^{\pm\frac{k_H}{k_y}\sqrt{2k_y N + C_y}} \tag{3.82}$$

with C_H another integration constant. We see that the Hubble expansion rate does not diverge at the separatrix, but it goes to the constant value C_H so that the energy density of the field remains finite. However, the derivative of the Hubble expansion rate near the separatrix evolves as

$$\dot{H} \simeq H^2 \frac{k_H}{\sqrt{2k_y N + C_y}} \tag{3.83}$$

so it goes to infinity as it approaches the separatrix. This signals a divergence in the pressure of the scalar field when the trajectory hits the separatrix so we find a future sudden singularity. This kind of singularity was first studied in Barrow (2004) and corresponds to the type II according to the classification performed in Nojiri et al. (2005).

So far, in our study we have focused on the case when only the π field contributes to the energy density of the universe and we have neglected any other possible component that might be present. We have shown that the only critical point is the pure vacuum Minkowski solution with $H = y = 0$. Moreover, we have shown that the separatrices can also act as attractors of the phase map and, when this happens, the evolution ends in a singularity where the derivative of the Hubble expansion rate diverges. In order to have a more realistic scenario, at least a matter component should be included. This will add a new dimension to the phase space and, thus, a new phenomenology is expected to arise. In particular, it could change some stability requirements and additional critical points might appear. Therefore, let us discuss in the following the case with matter fields.

If we include a pressureless matter component and use the variables H, $\hat{y}$ and[3] $\Omega_m \equiv \rho_m b_2/(6H^2)$, to describe the extended cosmological evolution, the corresponding autonomous system reads

$$\frac{d\hat{y}}{dN} = -\frac{\left(5 - 13\hat{y} + (9-41c_3)\hat{y}^2 + 57c_3\hat{y}^3 + 90c_3^2\hat{y}^4\right)\hat{y} + (1 - 3\hat{y} - 9c_3\hat{y}^2)H^2\Omega_m}{1 - 6\hat{y} + 6(1-5c_3)\hat{y}^2 + 52c_3\hat{y}^3 + 105c_3^2\hat{y}^4 - 2(1+6c_3\hat{y})H^2\Omega_m}$$
$$\frac{dH}{dN} = -\frac{2 - 8\hat{y} + (9-33c_3)\hat{y}^2 + 72c_3\hat{y}^3 + 135c_3^2\hat{y}^4 - 3(1+6c_3\hat{y})H^2\Omega_m}{1 - 6\hat{y} + 6(1-5c_3)\hat{y}^2 + 52c_3\hat{y}^3 + 105c_3^2\hat{y}^4 - 2(1+6c_3\hat{y})H^2\Omega_m} H$$
$$\frac{d\Omega_m}{dN} = \frac{1 + 2\hat{y} + 24c_3\hat{y}^2 - 12c_3\hat{y}^3 - 45c_3^2\hat{y}^4}{1 - 6\hat{y} + 6(1-5c_3)\hat{y}^2 + 52c_3\hat{y}^3 + 105c_3^2\hat{y}^4 - 2(1+6c_3\hat{y})H^2\Omega_m}\Omega_m. \quad (3.84)$$

Since we are seeking for critical points with $\Omega_m \neq 0$, we can solve for it from the vanishing of $d\hat{y}/dN$ to obtain the expression

$$\Omega_m H^2 = \frac{5 - 13\hat{y} + (9-41c_3)\hat{y}^2 + 57c_3\hat{y}^3 + 90c_3^2\hat{y}^4}{-1 + 3\hat{y} + 9c_3\hat{y}^2}\hat{y} \quad (3.85)$$

for the potential new critical points. Then, we can plug this relation into the remaining two equations given by the vanishing of dH/dN and $d\Omega_{md}/N$ to obtain the critical points. However, when doing so we end up with two equations that only depend on $\hat{y}$ which are, in general, incompatible for any value of c_3. Therefore, the inclusion of matter does not introduce new critical points in the phase map.

Above we have seen that the only critical point existing in the phase map of the Proxy theory (even if we include a dust component) is the vacuum Minkowski solution. As it is known, Gauss-Bonnet terms can give rise to accelerated expansion so that we will now modify our original Proxy theory by including a coupling of the scalar field to the Gauss-Bonnet term of the form $\mathcal{L}_{\pi GB} = a_4\pi(R_{\alpha\beta\gamma\delta}R^{\alpha\beta\gamma\delta} - 4R_{\alpha\beta}R^{\alpha\beta} + R^2)$. The cosmological equations in this case can be expressed as the following autonomous system:

$$\frac{d\hat{y}}{dN} = -\frac{\left(5 - 13\hat{y} + (9-41c_3)\hat{y}^2 + 57c_3\hat{y}^3 + 90c_3^2\hat{y}^4\right) + 4\epsilon\hat{H}^2(3 - 3\hat{y} - 10c_3\hat{y}^2)}{1 - 6\hat{y} + 6(1-5c_3)\hat{y}^2 + 52c_3\hat{y}^3 + 105c_3^2\hat{y}^4 + 16\hat{H}^4 + 8\epsilon\hat{H}^2(1 - 2\hat{y} - 9c_3\hat{y}^2)}\hat{y}$$
$$\frac{d\hat{H}}{dN} = -\frac{2 - 8\hat{y} + (9-33c_3)\hat{y}^2 + 72c_3\hat{y}^3 + 135c_3^2\hat{y}^4 + 16\hat{H}^4 + 12\epsilon\hat{H}^2(1 - 2\hat{y} + 8c_3\hat{y}^2)}{1 - 6\hat{y} + 6(1-5c_3)\hat{y}^2 + 52c_3\hat{y}^3 + 105c_3^2\hat{y}^4 + 16\hat{H}^4 + 8\epsilon\hat{H}^2(1 - 2\hat{y} - 9c_3\hat{y}^2)}\hat{H}$$

where $\epsilon \equiv \text{sign}(b_4)$ and $\hat{H} \equiv H\sqrt{|b_4|}$, with $b_4 \equiv a_4 M_p^3/\Lambda^3$ (and the number of e-folds is defined with such rescaled Hubble expansion rate). In order to look for critical points with $H \neq 0$, we solve for $\hat{H}^2$ from the equation $d\hat{y}/dN = 0$ and plug the obtained solution into the equation $d\hat{H}/dN = 0$. After doing so, we arrive at the following equation:

[3] Notice the factor b_2 in this definition of the matter density parameter that does not appear in the usual definition.

$$\frac{\hat{H}}{2 - 3\hat{y} - 10c_3\hat{y}^2} = 0 \tag{3.86}$$

whose solution is again $\hat{H} = 0$, signaling that the simple coupling of the scalar field to the Gauss-Bonnet term that we have considered is not able to introduce additional critical points.

To end, let us point out that the problematic term avoiding the existence of de Sitter critical points in the cosmological evolution is the πR term in the action. Our original approximation $\pi H \ll \dot{\pi}$ actually means that said term is negligible. However, our findings show that such a term cannot be consistently maintained small and it is the responsible for the absence of de Sitter solutions in the Proxy theory. Thus, a natural modification of it that will lead to de Sitter solutions consists in simply dropping the problematic term πR from the action. It is evident that this modified theory will have de Sitter solutions because in that case our previously used approximation is exact. In fact, such a term is the only one violating the shift symmetry so that without it, only the derivatives of the scalar field are physically relevant, but not the value of the field itself. We can proceed analogously as before to obtain the corresponding autonomous system and look for the critical points. When doing so, one can show that there are de Sitter critical points and that they are stable. In Fig. 3.6 we plot an example of the phase map for the case without the πR term in the action and we show the existence of the de Sitter attractor.

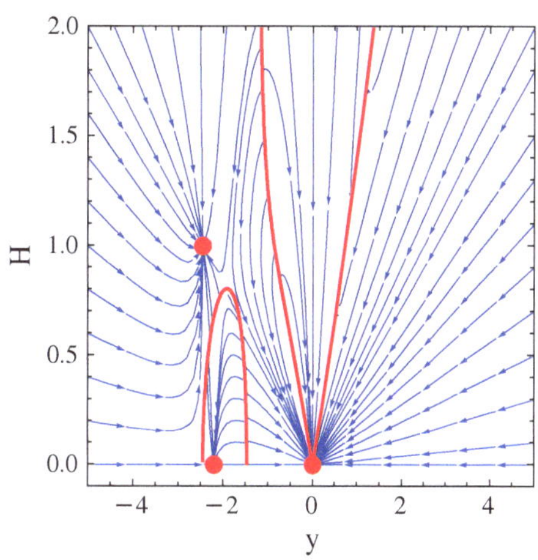

Fig. 3.6 In this figure we show an example of the phase map for the proxy theory without the πR term that spoils the existence of de Sitter critical points. We can see that the de Sitter solution is a an attractor of the cosmological evolution. The *red lines* denote the corresponding separatrices

3.7 Horndeski Vector Fields

We have seen that our proxy theory corresponds to a subclass of Horndeski scalar tensor interactions which contain second order equations of motion. The Galileon interactions present a subclass of Horndeski interactions as well. It has been shown that there is no theory equivalent to scalar Galileons for vector fields meaning that there is no "vector Galileons" besides the Maxwell kinetic term with second order equations of motion on flat space-times (Deffayet et al. 2013). However, motivated by the non-minimal coupling between the scalar field and Einstein tensor $G_{\mu\nu}$ and dual Riemann tensor $L_{\mu\nu\alpha\beta}$ we can ask a similar question for a vector field, i.e. what is the most general action for a vector field with a non-minimal coupling to gravity leading to second order equations of motion for both, the vector field and the gravitational sector. Hereby, we want preserve the gauge symmetry and so consider an action containing only kinetic terms for the vector field so that neither potential terms nor direct couplings of the vector field to curvature will be considered, but only derivative couplings. In principle, we could consider all possible terms involving a coupling of $F_{\mu\nu}$ to the Riemann tensor. However, in order to guarantee that the gravitational equations remain of second order, the couplings must be to a divergence-free tensor constructed with the Riemann tensor. We have seen in the above sections that in 4-dimensions, in addition to the metric there are only two tensors satisfying this condition, namely the Einstein tensor and the dual Riemann tensor defined by $L^{\alpha\beta\gamma\delta} = -\frac{1}{2}\epsilon^{\alpha\beta\mu\nu}\epsilon^{\gamma\delta\rho\sigma}R_{\mu\nu\rho\sigma}$, where $\epsilon_{\mu\nu\alpha\beta} = \epsilon_{[\mu\nu\alpha\beta]}$ is the Levi-Civita tensor. This divergenceless tensors are related to the non-trivial Lovelock invariants in 4 dimensions, i.e., the Ricci scalar and the Gauss-Bonnet term as explained in Sect. 1.3.

Since the Einstein tensor is symmetric, its contraction with $F_{\mu\nu}$ vanishes for symmetry reasons $G_{\mu\nu}F^{\mu\nu} = 0$. Note that even though $G^{\mu\nu}$ and $g^{\alpha\beta}$ are both divergence-free, their product $G^{\mu\nu}g^{\alpha\beta}$ is not so that we cannot allow the term $G^{\mu\nu}g^{\alpha\beta}F_{\mu\alpha}F_{\nu\beta}$. The same argument applies to the product of two Einstein tensors so that $G^{\mu\nu}G^{\alpha\beta}F_{\mu\alpha}F_{\nu\beta}$ is not allowed either. Only the dual Riemann tensor is divergence-free and it can be coupled to $F_{\mu\alpha}F_{\nu\beta}$. Thus, the desired action with only second order equations of motion is remarkably simple and reads

$$
\begin{aligned}
S &= \int d^4x\sqrt{-g}\left[-\frac{1}{2}M_p^2R - \frac{1}{4}F_{\mu\nu}F^{\mu\nu} + \frac{1}{4M^2}L^{\alpha\beta\gamma\delta}F_{\alpha\beta}F_{\gamma\delta}\right] \\
&= \int d^4x\sqrt{-g}\left[-\frac{1}{2}M_p^2R - \frac{1}{4}F_{\mu\nu}F^{\mu\nu}\right. \\
&\quad \left. + \frac{1}{2M^2}\left(RF_{\mu\nu}F^{\mu\nu} - 4R_{\mu\nu}F^{\mu\sigma}F^{\nu}{}_{\sigma} + R_{\mu\nu\alpha\beta}F^{\mu\nu}F^{\alpha\beta}\right)\right],
\end{aligned} \tag{3.87}
$$

where M^2 would be the only free parameter of the theory and its sign will be fixed by stability requirements. One might further wonder if interactions of the form $L^{\alpha\beta\gamma\delta}\tilde{F}_{\alpha\beta}\tilde{F}_{\gamma\delta}$ and $L^{\alpha\beta\gamma\delta}F_{\alpha\beta}\tilde{F}_{\gamma\delta}$ (with $\tilde{F}^{\alpha\beta} = \frac{1}{2}\epsilon^{\alpha\beta\mu\nu}F_{\mu\nu}$ being the dual of $F_{\mu\nu}$) could fulfill our requirements and be a valid interaction which should be taken into account. The latter one would explicitly break the parity invariance, but this symmetry is not one of our requirements. Nevertheless, at a closer look one realizes that these interactions are nothing else but coupling of the Riemann tensor to $F_{\mu\alpha}F_{\nu\beta}$. Since the Riemann tensor is not divergence-free, these interactions give rise to higher order equations of motion. So interactions of the form $L^{\alpha\beta\gamma\delta}F_{\alpha\beta}F_{\gamma\delta}$ fulfill our requirement of second order equation of motion and gauge invariance and are equivalent to $R^{\alpha\beta\gamma\delta}\tilde{F}_{\alpha\beta}\tilde{F}_{\gamma\delta}$, while interactions of the form $L^{\alpha\beta\gamma\delta}\tilde{F}_{\alpha\beta}\tilde{F}_{\gamma\delta}$ are not since they are equivalent to $R^{\alpha\beta\gamma\delta}F_{\alpha\beta}F_{\gamma\delta}$ and thus give higher order equations of motion. This reasoning can be understood from the fact that $F_{\mu\nu}$ is a closed form (so it satisfies Bianchi identities) whereas its dual is not. Furthermore, one might wonder whether the interaction changes if we contract the indices between the dual Riemann tensor and $F_{\mu\alpha}F_{\nu\beta}$ in a different way, i.e. whether $L^{\alpha\gamma\beta\delta}F_{\alpha\beta}F_{\gamma\delta}$ gives rise to a different interaction but it turns out to be that this term is proportional to $L^{\alpha\beta\gamma\delta}F_{\alpha\beta}F_{\gamma\delta}$ such that one can reabsorb its effect into M^2. Note also that this Horndeski interaction is the only interaction yielding second order equations of motion in four dimensions. In higher dimensions one can construct other non-minimal interactions based on the divergence-free tensors in that dimension; more precisely, one will have additional divergence-free tensors associated to the corresponding Lovelock invariants which can then be contracted with the field strength tensor.

After working out the scalar field case, Horndeski proved that this coupling is actually the only non-minimal coupling for the electromagnetic field leading to second order equations of motion and recovering Maxwell theory in flat spacetime. As can be seen from the action, Maxwell theory corresponds to the limit where the spacetime curvature is much smaller than M^2. In this limit, the non-minimal coupling is strongly suppressed with respect to the usual Maxwell term.

That the particular combination appearing in (3.87) leads to second order equations of motion can be easily understood. Only the non-minimal coupling contains more than two derivatives so this is the only term that could lead to higher order terms in the equations of motion. In order to show that the equations remain of second order is convenient to write the dual Riemann tensor as $L^{\alpha\beta\gamma\delta} = -\frac{1}{2}\epsilon^{\alpha\beta\mu\nu}\epsilon^{\gamma\delta\rho\sigma}R_{\mu\nu\rho\sigma}$. Then, it becomes apparent why the M^2-term will lead to second order equations of motion by virtue of the Bianchi identities for the Riemann tensor and $F_{\mu\nu}$. For instance, if we perform the variation of the non-minimal interaction with respect to A_μ, the only possible danger terms with higher order derivatives will come from derivatives applying on the Riemann tensor once we do integration by parts $\frac{1}{2}\epsilon^{\alpha\beta\mu\nu}\epsilon^{\gamma\delta\rho\sigma}\nabla_\gamma R_{\mu\nu\rho\sigma}F_{\alpha\beta}\delta A_\delta$. Using the Bianchi identity for the Riemann tensor $R_{\mu\nu[\rho\sigma;\gamma]} = 0$ we see that this dangerous terms automatically cancel.[4] We can also see this explicitly by computing the corresponding equations of motion. The non-gravitational field equations are therefore of second order and are given by

[4]The Bianchi identity for an arbitrary p-form $d\omega = 0$, where ω is the strength of the p-form, guaranties that the Lagrangian $\epsilon^{\mu\nu\cdots}\epsilon^{ab\cdots}\omega_{\mu\nu\ldots}\omega_{ab\ldots}\cdots(\partial_\rho\omega_{cd\ldots})\cdots(\partial_e\omega_{\sigma\tau\ldots})$ for the p-form will only give rise to second-order equations of motion (Deffayet et al. 2010).

$$\left[g^{\mu\rho}g^{\nu\sigma}-\frac{1}{M^2}L^{\mu\nu\rho\sigma}\right]\nabla_\nu F_{\rho\sigma}=0\,. \tag{3.88}$$

Because of the transversality of the dual Riemann tensor, we see that the above equation is divergence-free. Moreover, since $L^{\alpha\beta\gamma\delta}$ is divergence-free, we can also write the above equation as

$$\nabla_\nu\left[F^{\mu\nu}-\frac{1}{M^2}L^{\mu\nu\rho\sigma}F_{\rho\sigma}\right]=0, \tag{3.89}$$

which resembles the usual form of Maxwell equations.

Varying the action (3.87) with respect to the metric yields the following energy momentum tensor for the vector field

$$\begin{aligned}T_{\mu\nu}=&\frac{1}{2M^2}\left[-R_{\alpha\beta\gamma\delta}\tilde{F}^{\alpha\beta}\tilde{F}^{\gamma\delta}g_{\mu\nu}+2R_{\mu\beta\gamma\delta}\tilde{F}_\nu{}^\beta\tilde{F}^{\gamma\delta}+4\nabla^\gamma\nabla^\beta\left(\tilde{F}_{\mu\beta}\tilde{F}_{\gamma\nu}\right)\right]\\&-F_{\mu\alpha}F_\nu{}^\alpha+\frac{1}{4}g_{\mu\nu}F_{\alpha\beta}F^{\alpha\beta}\end{aligned} \tag{3.90}$$

Although the M^2-term of the energy momentum tensor might seem to contain more than second order derivatives, because $\tilde{F}^{\alpha\beta}$ is divergence-free in the absence of external currents, this is actually not the case. In fact, it can be written in the more suggestive form

$$\nabla_\gamma\nabla_\beta\left(\tilde{F}^{\mu\beta}\tilde{F}^{\gamma\nu}\right)=R^\mu_{\lambda\beta\gamma}\tilde{F}^{\lambda\beta}\tilde{F}^{\gamma\nu}+R_{\lambda\gamma}\tilde{F}^{\mu\lambda}\tilde{F}^{\gamma\nu}+\nabla_\gamma\tilde{F}^{\mu\beta}\nabla_\beta\tilde{F}^{\gamma\nu}\,. \tag{3.91}$$

If a current is present so that $\tilde{F}^{\alpha\beta}$ is no longer divergence-free, this equation acquires a contribution from the current. In Eq. (3.91) one sees explicitly that only second derivatives are present. For more detail see our work in Jiménez et al. (2013).

If one gives up on the gauge invariance meaning that we allow for terms which are not invariant under $A_\mu\to A_\mu+\partial_\mu\theta$ then one can indeed construct "vector Galileons" or in more general Horndeski vector interactions giving rise to three propagating degrees of freedom with second order equations of motion. In the following our purpose will be to find a generalization of the Proca action for a massive vector field. In particular, we aim to finding the general lagrangian terms (including derivative non-linear interactions) with the requirement that only three physical degrees of freedom propagate, as it corresponds to a general massive spin-1 field. We will first analyse the case of a flat Minkowksi spacetime and we will generalize our results to the case of an arbitrary curved spacetime. This is an unpublished result which we quote here without entering into much details. The Proca action is the theory describing a massive vector field, which propagates the corresponding 3 polarizations (2 transverse plus 1 longitudinal) . The mass term breaks explicitly the $\mathcal{U}(1)$ gauge invariance such that the longitudinal mode propagates as well. However, the zero component of the vector field does not propagate. The Proca action is given by

$$\mathcal{S}_{\text{Proca}} = \int d^4x \left[-\frac{1}{4} F_{\mu\nu}^2 + \frac{1}{2} m^2 A^2 \right] \tag{3.92}$$

In this theory, the temporal component of the vector field does not propagate and generates a primary constraint. The consistency condition of this primary constraint generates a secondary constraint, whose Poisson bracket with the primary constraint is proportional to the mass so that only in the massless case it corresponds to a first class constraint generating a gauge symmetry.

Now we want to generalize the above Proca action to include derivative self-interactions of the vector field, but without changing the number of propagating degrees of freedom. In order to obtain such interactions, we will analyze all the possible Lorentz invariant terms that can be built at each order. The simplest modification is of course promoting the mass term to an arbitrary potential $V(A^2)$, since this trivially does not modify the number of degrees of freedom.

The first term that we can have to the next order in the vector field is simply

$$\mathcal{L}_3 = f_3(A^2)(\partial \cdot A) \tag{3.93}$$

with $f_3(A^2)$ and arbitrary function of the vector field norm. It is a trivial observation that in (3.93) the component A_0 does not propagate, even if we include the Maxwell kinetic term, and it acts as a lagrange multiplier. The easiest way to see it is by computing the corresponding Hessian, which vanishes trivially. Also notice that the presence of the function f_3 is crucial since if it was simply a constant, that term would be a total divergence and, thus, with no contribution to the field equations.

To next order, the independent interaction terms that we can have are given by

$$\mathcal{L}_4 = f_4(A^2)\left[c_1(\partial \cdot A)^2 + c_2 \partial_\rho A_\sigma \partial^\rho A^\sigma + c_3 \partial_\rho A_\sigma \partial^\sigma A^\rho\right] \tag{3.94}$$

with a priori free parameters c_1, c_2 and c_3 and f_4 an arbitrary function. Now, we need to fix the parameters such that only three physical degrees of freedom propagate, i.e., such that we still have a second class constraint. In order to eliminate one propagating degree of freedom, we need a constraint equation, which is guaranteed if the determinant of the Hessian matrix vanishes. The Hessian matrix for (3.94) is given by

$$H^{\mu\nu}_{\mathcal{L}_4} = \frac{\partial^2 \mathcal{L}_4}{\partial \dot{A}_\mu \partial \dot{A}_\nu} = f(A^2) \begin{pmatrix} 2(c_1+c_2+c_3) & 0 & 0 & 0 \\ 0 & -2c_2 & 0 & 0 \\ 0 & 0 & -2c_2 & 0 \\ 0 & 0 & 0 & -2c_2 \end{pmatrix} \tag{3.95}$$

For a vanishing Hessian matrix we have two possibilities. First possibility corresponds to choosing $c_2 = 0$. In this case the Hessian matrix contains three vanishing eigenvalues corresponding to three constraints. Therefore, if we choose $c_2 = 0$, only the zero component of the vector field propagate while the other three degrees of freedom do not propagate. The other possibility for a vanishing determinant of the Hessian matrix corresponds to $c_1 + c_2 + c_3 = 0$. Without loss of generality we can set $c_1 = 1$ and therefore $c_3 = -(1 + c_2)$. In this case the Hessian matrix only contains one vanishing eigenvalue and hence only one propagating constraint. This case corresponds to three propagating degrees of freedom.

$$\mathcal{L}_4 = f_4(A^2)\left[(\partial \cdot A)^2 + c_2\partial_\rho A_\sigma \partial^\rho A^\sigma - (1 + c_2)\partial_\rho A_\sigma \partial^\sigma A^\rho\right] \tag{3.96}$$

The vanishing of the determinant of the Hessian matrix guaranties the existence of a constraint. To find the expression for the constraint, we have to compute the conjugate momentum $\Pi^\mu_{\mathcal{L}_4} = \frac{\partial \mathcal{L}_4}{\partial \dot{A}_\mu}$. The zero component of the conjugate momentum is given by

$$\Pi^0_{\mathcal{L}_4} = -2f_4(A^2)\vec{\nabla}\vec{A} \tag{3.97}$$

As one can see, the zero component of the conjugate momentum does not contain any time derivative yielding the constraint equation.

For the next order interactions the possible terms are the following:

$$\begin{aligned}\mathcal{L}_5 = f_5(A^2)\ \Big[&d_1(\partial \cdot A)^3 - 3d_2(\partial \cdot A)\partial_\rho A_\sigma \partial^\rho A^\sigma - 3d_3(\partial \cdot A)\partial_\rho A_\sigma \partial^\sigma A^\rho \\ &+ 2d_4\partial_\rho A_\sigma \partial^\gamma A^\rho \partial^\sigma A_\gamma + 2d_5\partial_\rho A_\sigma \partial^\gamma A^\rho \partial_\gamma A^\sigma\Big]\end{aligned} \tag{3.98}$$

with the free parameters d_1, d_2, d_3, d_4 and d_5. In this quantic Lagrangian (3.98) the additional possible term $\partial_\sigma A_\rho \partial^\gamma A^\rho \partial_\gamma A^\sigma$ is equal to $\partial_\rho A_\sigma \partial^\gamma A^\rho \partial_\gamma A^\sigma$ since $\partial^\gamma A^\rho \partial_\gamma A^\sigma$ is symmetric under the exchange of ρ and σ. The Hessian matrix for this quintic Lagrangian is giving by

$$\begin{aligned}
H^{00}_{\mathcal{L}_5} &= -6(d_1 - d_2 - d_3)(\vec{\nabla}\vec{A}) + 6(d_1 - 3d_2 - 3d_3 + 2(d_4 + d_5))\dot{A}_t \\
H^{01}_{\mathcal{L}_5} &= H^{10}_{\mathcal{L}_5} = (6d_3 - 2(3d_4 + d_5))A_{t,x} + 2(3d_2 - 2d_4)A_{x,t} \\
H^{02}_{\mathcal{L}_5} &= H^{20}_{\mathcal{L}_5} = (6d_3 - 2(3d_4 + d_5))A_{t,y} + 2(3d_2 - 2d_4)A_{y,t} \\
H^{03}_{\mathcal{L}_5} &= H^{30}_{\mathcal{L}_5} = (6d_3 - 2(3d_4 + d_5))A_{t,z} + 2(3d_2 - 2d_4)A_{z,t} \\
H^{11}_{\mathcal{L}_5} &= -6d_2A^2_\alpha(A_{z,z} + A_{y,y}) - 2(3d_2 - 2d_4)(A_{x,x} - A_{t,t}) \\
H^{12}_{\mathcal{L}_5} &= H^{21}_{\mathcal{L}_5} = 2d_5(A_{x,y} + A_{y,x}) \\
H^{13}_{\mathcal{L}_5} &= H^{31}_{\mathcal{L}_5} = 2d_5(A_{x,z} + A_{z,x}) \\
H^{22}_{\mathcal{L}_5} &= 2(-3d_2A_{z,z} + (-3d_2 + 2d_5)A_{y,y} - 3d_2A_{x,x} + (3d_2 - 2d_5)A_{t,t}) \\
H^{23}_{\mathcal{L}_5} &= H^{32}_{\mathcal{L}_5} = 2d_5(A_{y,z} + A_{z,y}) \\
H^{33}_{\mathcal{L}_5} &= (-6d_2 + 4d_5)A_{z,z} - 6d_2(A_{y,y} + A_{x,x}) + 2(3d_2 - 2d_5)A_{t,t}
\end{aligned} \tag{3.99}$$

In order to have only three propagating degrees of freedom the parameters need to fulfill the following conditions

$$\begin{aligned} d_1 - d_2 - d_3 &= 0, \qquad & d_1 - 3d_2 - 3d_3 + 2(d_4 + d_5) &= 0, \\ 3d_3 - 3d_4 - d_5 &= 0, \qquad & 3d_2 - 2d_5 &= 0 \end{aligned} \tag{3.100}$$

which are fulfilled by choosing (again without loss of generality we can choose $d_1 = 1$)

$$d_3 = 1 - d_2, \qquad d_4 = 1 - \frac{3d_2}{2}, \qquad d_5 = \frac{3d_2}{2} \tag{3.101}$$

Hence, the quintic Lagrangian with only three physical propagating degrees of freedom is given by

$$\begin{aligned} \mathcal{L}_5 = f_5(A^2)\Big[&(\partial \cdot A)^3 - 3d_2(\partial \cdot A)\partial_\rho A_\sigma \partial^\rho A^\sigma - 3(1-d_2)(\partial \cdot A)\partial_\rho A_\sigma \partial^\sigma A^\rho \\ &+ 2\left(1 - \frac{3d_2}{2}\right)\partial_\rho A_\sigma \partial^\gamma A^\rho \partial^\sigma A_\gamma + 2\left(\frac{3d_2}{2}\right)\partial_\rho A_\sigma \partial^\gamma A^\rho \partial_\gamma A^\sigma\Big] \end{aligned} \tag{3.102}$$

The Hessian matrix with this chosen parameters then becomes

$$H^{\mu\nu}_{\mathcal{L}_5} = f_5(A^2)\begin{pmatrix} 0 & 0 & 0 & 0 \\ 0 & -6d_2(A_{z,z} + A_{y,y}) & 3d_2(A_{x,y} + A_{y,x}) & 3d_2(A_{x,z} + A_{z,x}) \\ 0 & 3d_2(A_{x,y} + A_{y,x}) & -6d_2(A_{z,z} + A_{x,x}) & 3d_2(A_{y,z} + A_{z,y}) \\ 0 & 3d_2(A_{x,z} + A_{z,x}) & 3d_2(A_{y,z} + A_{z,y}) & -6d_2(A_{y,y} + A_{x,x}) \end{pmatrix}$$

with a vanishing determinant $\det(H^{\mu\nu}_{\mathcal{L}_5}) = 0$ as we required. The corresponding zero component of the conjugate momentum $\Pi^\mu_{\mathcal{L}_5} = \frac{\partial \mathcal{L}_5}{\partial \dot{A}_\mu}$ is given by

$$\begin{aligned} \Pi^0_{\mathcal{L}_5} = -3f_5(A^2)\Big(&d_2(A^2_{x,z} + A^2_{y,z} + A^2_{x,y}) - 2A_{z,z}A_{y,z} - 2(-1+d_2)A_{y,z}A_{z,y} \\ &+ d_2A^2_{z,y} - 2(A_{z,z} + A_{y,y})A_{x,x} + 2A_{x,y}A_{y,x} - 2d_2A_{x,y}A_{y,x} + d_2A^2_{y,x} \\ &- 2(-1+d_2)A_{x,z}A_{z,x} + d_2A^2_{z,x}\Big) \end{aligned} \tag{3.103}$$

As you can see, there is no time derivatives appearing in the expression of the zero component of the conjugate momentum, representing the constraint equation. Thus the Lagrangian for the most general Proca vector field yields

$$\begin{aligned} \mathcal{L}_{\text{Proca}} &= -\frac{1}{4}F^2_{\mu\nu} + \frac{1}{2}m^2(A_\mu)^2 \\ \mathcal{L}_3 &= f_3(A^2)(\partial \cdot A) \\ \mathcal{L}_4 &= f_4(A^2)\left[(\partial \cdot A)^2 + c_2\partial_\rho A_\sigma \partial^\rho A^\sigma - (1+c_2)\partial_\rho A_\sigma \partial^\sigma A^\rho\right] \end{aligned}$$

$$\mathcal{L}_5 = f_5(A^2)\Big[(\partial\cdot A)^3 - 3d_2(\partial\cdot A)\partial_\rho A_\sigma\partial^\rho A^\sigma - 3(1-d_2)(\partial\cdot A)\partial_\rho A_\sigma\partial^\sigma A^\rho$$
$$+2\left(1-\frac{3d_2}{2}\right)\partial_\rho A_\sigma\partial^\gamma A^\rho\partial^\sigma A_\gamma + 2\left(\frac{3d_2}{2}\right)\partial_\rho A_\sigma\partial^\gamma A^\rho\partial_\gamma A^\sigma\Big] \quad (3.104)$$

The interactions can be also expressed in terms of the Levi-Civita tensors

$$\mathcal{L}_2 = -\frac{1}{24}\mathcal{E}^{\mu\nu\alpha\beta}\mathcal{E}_{\mu\nu\alpha\beta}f_2(A^2) = f_2(A^2) \quad (3.105)$$

$$\mathcal{L}_3 = -\frac{1}{6}\mathcal{E}^{\mu\nu\alpha\beta}\mathcal{E}^{\rho}{}_{\nu\alpha\beta}f_3(A^2)\partial_\mu A_\rho = f_3(A^2)(\partial\cdot A) \quad (3.106)$$

$$\mathcal{L}_4 = -\frac{1}{2}\left(\mathcal{E}^{\mu\nu\alpha\beta}\mathcal{E}^{\rho\sigma}{}_{\alpha\beta}f_4(A^2)\partial_\mu A_\rho\partial_\nu A_\sigma + c_2\mathcal{E}^{\mu\nu\alpha\beta}\mathcal{E}^{\rho\sigma}{}_{\alpha\beta}f_4(A^2)\partial_\mu A_\nu\partial_\rho A_\sigma\right)$$
$$= f_4(A^2)\left[(\partial\cdot A)^2 + c_2\partial_\rho A_\sigma\partial^\rho A^\sigma - (1+c_2)\partial_\rho A_\sigma\partial^\sigma A^\rho\right] \quad (3.107)$$

$$\mathcal{L}_5 = -f_5(A^2)((1-\frac{3}{2}d_2)\mathcal{E}^{\mu\nu\alpha\beta}\mathcal{E}^{\rho\sigma\gamma}{}_{\beta}\partial_\mu A_\rho\partial_\nu A_\sigma\partial_\alpha A_\gamma$$
$$+\frac{3}{2}d_2(\mathcal{E}^{\mu\nu\alpha\beta}\mathcal{E}^{\rho\sigma\gamma}{}_{\beta}\partial_\mu A_\rho\partial_\nu A_\sigma\partial_\gamma A_\alpha))$$
$$= f_5(A^2)\Big[(\partial\cdot A)^3 - 3d_2(\partial\cdot A)\partial_\rho A_\sigma\partial^\rho A^\sigma - 3(1-d_2)(\partial\cdot A)\partial_\rho A_\sigma\partial^\sigma A^\rho$$
$$+2\left(1-\frac{3d_2}{2}\right)\partial_\rho A_\sigma\partial^\gamma A^\rho\partial^\sigma A_\gamma + 2\left(\frac{3d_2}{2}\right)\partial_\rho A_\sigma\partial^\gamma A^\rho\partial_\gamma A^\sigma\Big] \quad (3.108)$$

The Lagrangians $\mathcal{L}_2$, $\mathcal{L}_3$, $\mathcal{L}_4$, $\mathcal{L}_5$ in (3.104) are the only Vector interactions which propagate three degrees of freedom. The interactions beyond the quintic order give zero contributions, hence the serie stops here.

$$\mathcal{L}_6 = \mathcal{E}^{\mu\nu\alpha\beta}\mathcal{E}^{\rho\sigma\delta\gamma}A_\kappa A^\kappa\partial_\mu A_\rho\partial_\nu A_\sigma\partial_\alpha A_\delta\partial_\beta A_\gamma = 0 \quad (3.109)$$

When we constructed the interactions out of the Levi-Civita tensors, the indices of the potential interactions were always contracted with each other, meaning that we had always $A_\mu A^\mu$. One might wonder, if it yields different interactions once the indices of the mass term are contracted with the Levi-Civita tensors as well. But on closer inspection one can see that they give rise to the same interactions once integrations by part are performed (where we choosed without loss of generality $f_{2,3,4,5}(A^2) = (A_\mu)^2$)

$$\mathcal{L}_2^{al} = -\frac{1}{6}\mathcal{E}^{\mu\nu\alpha\beta}\mathcal{E}^{\rho}{}_{\nu\alpha\beta}A_\mu A_\rho = (A_\mu)^2 \quad (3.110)$$

$$\mathcal{L}_3^{al} = -\frac{1}{2}\mathcal{E}^{\mu\nu\alpha\beta}\mathcal{E}^{\rho\sigma}{}_{\alpha\beta}A_\mu A_\rho\partial_\nu A_\sigma = (A_\mu)^2(\partial\cdot A) - A^\mu A^\nu\partial_\nu A_\mu \quad (3.111)$$

$$\mathcal{L}_4^{al} = -\mathcal{E}^{\mu\nu\alpha\beta}\mathcal{E}^{\rho\sigma\delta}{}_{\beta}A_\mu A_\rho\partial_\nu A_\sigma\partial_\alpha A_\delta = (A_\mu)^2\left[(\partial\cdot A)^2 - \partial_\rho A_\sigma\partial^\sigma A^\rho\right]$$

$$- 2A^{\mu}A^{\nu}\partial_{\nu}A^{\mu}(\partial\cdot A) + 2A\mu A^{\nu}\partial_{\nu}A_{\rho}\partial^{\rho}A_{\mu} \tag{3.112}$$

$$\mathcal{L}_5^{\text{al}} = \mathcal{E}^{\mu\nu\alpha\beta}\mathcal{E}^{\rho\sigma\delta\gamma}A_{\mu}A_{\rho}\partial_{\nu}A_{\sigma}\partial_{\alpha}A_{\delta}\partial_{\beta}A_{\gamma} = (A_{\mu})^2\Big[-(\partial\cdot A)^3 + 3(\partial\cdot A)\partial_{\rho}A_{\sigma}\partial^{\sigma}A^{\rho}$$
$$2\partial_{\rho}A_{\sigma}\partial^{\gamma}A^{\rho}\partial^{\sigma}A_{\gamma}\Big] + 3A^{\mu}A^{\nu}\partial_{\nu}A_{\mu}(\partial\cdot A)^2 - 6A^{\mu}A^{\nu}\partial_{\nu}A_{\rho}\partial^{\rho}A_{\mu}(\partial\cdot A)$$
$$+ 6A^{\mu}A^{\nu}\partial_{\nu}A_{\rho}\partial^{\rho}A_{\gamma}\partial^{\gamma}A_{\mu} - 3A^{\mu}A^{\nu}\partial_{\nu}A_{\mu}\partial_{\rho}A_{\sigma}\partial^{\sigma}A^{\rho} \tag{3.113}$$

Indeed, let us pay more attention to the special case with $f_{2,3,4,5}(A^2) = (A_{\mu})^2$

$$\mathcal{L}_{\text{Proca}} = -\frac{1}{4}F_{\mu\nu}^2 + \frac{1}{2}m^2(A_{\mu})^2$$
$$\mathcal{L}_3 = (A_{\mu})^2(\partial\cdot A)$$
$$\mathcal{L}_4 = (A_{\mu})^2\Big[(\partial\cdot A)^2 + c_2\partial_{\rho}A_{\sigma}\partial^{\rho}A^{\sigma} - (1+c_2)\partial_{\rho}A_{\sigma}\partial^{\sigma}A^{\rho}\Big]$$
$$\mathcal{L}_5 = (A_{\mu})^2\Big[(\partial\cdot A)^3 - 3d_2(\partial\cdot A)\partial_{\rho}A_{\sigma}\partial^{\rho}A^{\sigma} - 3(1-d_2)(\partial\cdot A)\partial_{\rho}A_{\sigma}\partial^{\sigma}A^{\rho}$$
$$+2\left(1-\frac{3d_2}{2}\right)\partial_{\rho}A_{\sigma}\partial^{\gamma}A^{\rho}\partial^{\sigma}A_{\gamma} + 2\left(\frac{3d_2}{2}\right)\partial_{\rho}A_{\sigma}\partial^{\gamma}A^{\rho}\partial_{\gamma}A^{\sigma}\Big] \tag{3.114}$$

Now, if we extract out only the longitudinal mode of the vector field, meaning that we replace the vector field as $A_{\mu} = \partial_{\mu}\phi$ then we recover the Galileon interactions

$$\mathcal{L}_2 = (\partial\pi)^2$$
$$\mathcal{L}_3 = (\partial\pi)^2\Box\pi$$
$$\mathcal{L}_4 = (\partial\pi)^2\Big[(\Box\pi)^2 - (\partial_{\mu}\partial_{\nu}\pi)^2\Big]$$
$$\mathcal{L}_5 = (\partial\pi)^2\Big[(\Box\pi)^3 - 3\Box\pi(\partial_{\mu}\partial_{\nu}\pi)^2 + 2(\partial_{\mu}\partial_{\nu}\pi)^3\Big] \tag{3.115}$$

This is another way of showing that the interactions we found for the vector field indeed only propagate three degrees of freedom, since when we plug in the longitudinal mode, we obtain the Galileon interaction with at most second order equations of motion. The terms for the vector field $\partial_{\rho}A_{\sigma}\partial^{\rho}A^{\sigma}$ and $\partial_{\rho}A_{\sigma}\partial^{\sigma}A^{\rho}$ are not the same, but when we replace $A_{\mu} = \partial_{\mu}\phi$, they are since the derivatives acting on the scalar field commute $\partial_{\mu}\partial_{\nu}\pi = \partial_{\nu}\partial_{\mu}\pi$. This has a huge concequence: the interactions for the vector field have more free parameters than the Galileon interactions. It means that if we had started with the Galileon interactions and performed the replacement $\partial_{\mu}\phi \to A_{\mu}$ we would have been missing some of the interactions which also yield three propagating degrees of freedom. The vector interaction have two more free parameters (namely what we called c_2 and d_2 in (3.114)). In fact, an alternative way of finding our generalized Proca action is by restoring the U(1) gauge invariance and imposing that the Stueckelberg field propagates only one degree of freedom, i.e., it satisfies second order field equations. One must be careful though, since in addition to the pure Stueckelberg sector, it is also necessarily to analyse the terms mixing the Stueckelberg field and the vector field. For L_4, no additional constraints arise from the mixing terms, since we obtain terms of the general form $K^{\mu\nu}(A_\mu)\partial_\mu\phi\partial_\nu\phi$, which automatically leads to second order contributions for ϕ. However, for L_5 we obtain terms like $K^{\alpha\beta\gamma\delta}(A_\mu)\partial_\alpha\partial_\beta\phi\partial_\gamma\partial_\delta\phi$ so we need to impose the tensor $K^{\alpha\beta\gamma\delta}(A_\mu)$ to have the correct structure. It is also worth noticing that the arbitrary functions appearing in our generalized Proca action have been assumed to be functions of A^2. In the Stueckelberg language, this is so in order to guarantee the second order nature of the field equations with respect to ϕ. There are however additional contribution upon which those arbitrary functions might depend without altering the number of degrees of freedom. Such terms are those for which the Stueckelberg field give a trivial contribution, i.e., those which are U(1) gauge invariant. Thus, those functions could actually depend also on the combinations F^2 or FF^*. From the vector field perspective, these terms do not contain time derivatives of A_0, so that it will not spoil the existence of the constraints. One must be cautious however, since arbitrary functions of such invariants typically give rise to violations of the hyperbolicity of the field equations. The equations of motion for (3.114) are given by

$$
\begin{aligned}
\mathcal{E}_{\rm Proca} &= \nabla^\nu F_{\mu\nu} + m^2 A_\mu \\
\mathcal{E}_3 &= 2A_\mu(\partial\cdot A) - 2A^\nu\partial_\mu A_\nu \\
\mathcal{E}_4 &= 2\Big(A_\mu\Big[(\partial\cdot A)^2 - (1+c_2)\partial_\rho A_\sigma\partial^\sigma A^\rho + c_2\partial_\rho A_\sigma\partial^\sigma A^\rho\Big] \\
&\quad + c_2 A^2(-\Box A_\mu + \partial_\nu\partial_\mu A^\nu) - 2c_2 A^\rho\partial_\nu A_\rho\partial^\nu A_\mu - 2(\partial\cdot A)A^\rho\partial_\mu A_\rho \\
&\quad + 2(1+c_2)A^\rho\partial_\nu A_\rho\partial_\mu A^\nu\Big) \\
\mathcal{E}_5 &= 2A_\mu\Big[(\partial\cdot A)^3 + 3(-1+d_2)(\partial\cdot A)\partial_\rho A_\sigma\partial^\sigma A^\rho - 3d_2(\partial\cdot A)\partial_\rho A_\sigma\partial^\rho A^\sigma \\
&\quad + (2-3d_2)\partial_\rho A_\sigma\partial^\gamma A^\rho\partial^\sigma A_\gamma + 3d_2\partial_\rho A^\sigma\partial^\rho A^\gamma\partial_\sigma A_\gamma\Big] \\
&\quad - 3A^\rho\Big(-d_2(4\partial_\nu A_\rho\partial^\nu A_\mu(\partial\cdot A) - 2(\partial_\nu A_\mu\partial^\nu A_\sigma + \partial^\nu A_\mu\partial_\sigma A_\nu)\partial^\sigma A_\rho \\
&\quad + A_\rho(\partial^\nu A_\mu(\partial_\sigma\partial_\nu A^\sigma - \Box A_\nu) + 2(\partial\cdot A)(\Box A_\mu - \partial_\sigma\partial_\mu A^\sigma) \\
&\quad + (\partial_\nu\partial_\mu A_\sigma - 2\partial_\sigma\partial_\nu A_\mu + \partial_\sigma\partial_\mu A_\nu)\partial^\sigma A^\nu)) \\
&\quad + 2((\partial\cdot A)^2 + ((-1+d_2)\partial_\nu A_\sigma - d_2\partial_\sigma A_\nu)\partial^\sigma A^\nu)\partial_\mu A_\rho \\
&\quad + (4(-1+d_2)\partial_\nu A_\rho(\partial\cdot A) + d_2 A_\rho(-\partial_\sigma\partial_\nu A^\sigma + \Box A_\nu) \\
&\quad + 2((2-3d_2)\partial_\nu A_\sigma + d_2\partial_\sigma A_\nu)\partial^\sigma A_\rho)\partial_\mu A^\nu\Big) \qquad (3.116)
\end{aligned}
$$

Note also that the equations of motion for the vector field does reproduce the equations of motion of the Galileon if we take the divergence of it and replace $A_\mu = \partial_\mu\phi$.

Now we would like to consider the interactions on a curved space-time. The Lagrangian in curved space-time becomes (with the short cut $X = -\frac{1}{2}A_\mu^2$)

$$
\begin{aligned}
\mathcal{L}_{\rm kin} &= -\frac{1}{4}F_{\mu\nu}^2 \\
\mathcal{L}_2 &= G_2(X) \\
\mathcal{L}_3 &= G_3(X)(D_\mu A^\mu) \\
\mathcal{L}_4 &= G_4(X)R + G_{4,X}\Big[(D_\mu A^\mu)^2 + c_2 D_\rho A_\sigma D^\sigma A^\rho - (1+c_2) D_\rho A_\sigma D^\sigma A^\rho\Big] \\
\mathcal{L}_5 &= G_5(X) G_{\mu\nu} D^\mu A^\nu - \frac{1}{6} G_{5,X}\Big[(D_\mu A^\mu)^3 - 3d_2 (D_\mu A^\mu) D_\rho A_\sigma D^\rho A^\sigma \\
&\quad - 3(1-d_2)(D_\mu A^\mu) D_\rho A_\sigma D^\sigma A^\rho + 2(1-\frac{3d_2}{2}) D_\rho A_\sigma D^\gamma A^\rho D^\sigma A_\gamma \\
&\quad + 2(\frac{3d_2}{2}) D_\rho A_\sigma D^\gamma A^\rho D_\gamma A^\sigma\Big] \\
&\quad + e_1 G^{\mu\nu} A_\mu A_\nu + e_2 L^{\mu\nu\alpha\beta} F_{\mu\nu} F_{\alpha\beta}
\end{aligned}
\tag{3.117}
$$

All these interaction give only rise to three propagating degrees of freedom in curved background.

3.8 Summary and Discussion

In this chapter we have started with the decoupling limit interactions of massive gravity and constructed a proxy theory to it by covariantizing the interactions. The resulting theory represents a specific class of non-minimally coupled scalar-tensor interactions. We have shown that the equations of motion contain at most second order derivatives acting on the fields. We studied the theory in the context of cosmology and were able to find self-accelerating solutions in a given approximation. However, beyond this approximation we have shown that self-accelerating solutions do not exist due to the existence of the πR coupling. But once one is willing to give up on this term, one can successfully construct stable de Sitter solutions. This would still correspond to a subclass of Horndeski theories and in their own right still be worth to investigate in more detail for cosmology.

On solar system and galactic scales gravity is very well compatible with General Relativity and correction terms from a modified gravity model like the Proxy theory need to be investigated carefully. Since we constructed the Proxy theory from the decoupling limit of massive gravity, the effects of Proxy theory on these scales are cloaked by the Vainshtein mechanism exactly in the same way as in massive gravity, where the scalar field interactions become appreciable to freeze out the field fluctuations, yet with some observational signatures on larger scales in cosmic structure formation. In contrast to the decoupling limit the self-accelerating solution we found in the approximated regime differs from a ΛCDM cosmological evolution. The alteration of the Hubble function can be experimentally tested using the distance-redshift relation of supernovae, and measurements of the angular diameter distance as a function of redshift, as in the case of the cosmic microwave background and the baryon acoustic oscillations.

Apart from these geometrical tests, the time-dependence of evolving cosmic structures can be investigated, and the influence of the gravitational theory on the geodesics of relativistic (photons) and nonrelativistic (dark matter) test particles. The first category includes gravitational lensing and the Sachs-Wolfe effects, which have been shown to differ from their GR-expectation in some modified gravity theories, and a similar result can be expected in Proxy theory.

The second category includes the homogeneous growth of the cosmic structure, and the formation of galaxies and clusters of galaxies by gravitational collapse. Again, the additional scalar degree of freedom would influence the time sequence of gravitational clustering and the evolution of peculiar velocities, as well as the number density of collapsed objects. In particular, we expect that it would enhance gravitational clustering since the collapse threshold for density fluctuations in the large-scale structure would be lowered, leading to a higher comoving number density of galaxies and clusters of galaxies. Naturally, these changes are degenerate with a different choice of cosmological parameters and with introducing non-Gaussian initial conditions, which would be very interesting to quantify.

Recent discrepancies of ΛCDM with observational data on large scales include the number of very massive clusters, the strong lensing cross section, anomalous multipole moments of the CMB, the axis of evil, and large peculiar velocities. It is beyond the scope of this thesis to address these issues using the Proxy theory, but we propose how to proceed from constructing a Proxy theory to providing observationally testable quantities. From our point of view it is advisable to focus on probes of large scales, due to difficulties related to nonlinear structure formation and the influence of baryons on small scales. Natural questions concern the homogeneous dynamics of the Universe, the formation of structures and the shape of geodesics of relativistic and nonrelativistic particles. Basically, using our proxy theory one should be able to make predictions concerning these four issues and the combination of the four should give insight into the nature of the gravitational sector. In our proxy theory, from the modified field equation the Hubble-function can be derived easily, which allows the definition of distance measures, needed in the interpretation of supernova data. Cosmic structure growth tests the Newtonian limit for slowly-moving particles and describes the clustering of galaxies and the growth of structures which are investigated by e.g. gravitational light deflection. Lensing, in turn, makes use of the geodesic equation for relativistic particles, and measures the correlation function of the matter density, weighted with the lensing efficiency function, which in turn is derived from distance measures. It would be quite interesting to study these observational consequences and constraints of our proxy theory in future works. An additional complication comes from the cosmological application of the Vainshtein mechanism which works such that at early times, the scalar field interactions are dominated by self-interaction, which suppresses their energy density relative to that of matter or radiation. If the matter density has decreased sufficiently by cosmic expansion, the scalar field constitutes an important contribution to the energy density of the universe and drives cosmic expansion.

References

Barrow JD (2004) Sudden future singularities. Class Quant Grav 21:L79–L82. doi:10.1088/0264-9381/21/11/L03

Charmousis C, Copeland EJ, Padilla A, Saffin PM (2012) General second order scalar-tensor theory, self tuning, and the Fab Four. Phys Rev Lett 108:051101. doi:10.1103/PhysRevLett.108.051101

de Rham C, Heisenberg L (2011) Cosmology of the galileon from massive gravity. Phys Rev D 84:043503. doi:10.1103/PhysRevD.84.043503

de Rham C, Tolley AJ (2010) DBI and the galileon reunited JCAP 1005:015. doi:10.1088/1475-7516/2010/05/015

Deffayet C, Esposito-Farese G, Vikman A (2009) Covariant Galileon Phys Rev D 79:084003. doi:10.1103/PhysRevD.79.084003

Deffayet C, Deser S, Esposito-Farese G (2010) Arbitrary p-form Galileons. Phys Rev D 82:061501. doi:10.1103/PhysRevD.82.061501

Deffayet C, Gumrukcuoglu AE, Mukohyama S, Yi W (2013) A no-go theorem for generalized vector Galileons on flat spacetime

Horndeski GW (1974) Second-order scalar-tensor field equations in a four-dimensional space. Int J Theor Phys 10:363–384. doi:10.1007/BF01807638

Jiménez JB, Durrer R, Heisenberg L, Thorsrud M (2013) Stability of Horndeski vector-tensor interactions. JCAP 1310:064. doi:10.1088/1475-7516/2013/10/064

Kimura R, Yamamoto K (2012) Constraints on general second-order scalar-tensor models from gravitational Cherenkov radiation. JCAP 1207:050. doi:10.1088/1475-7516/2012/07/050

Nojiri S, Odintsov SD, Tsujikawa S (2005) Properties of singularities in (phantom) dark energy universe. Phys Rev D 71:063004. doi:10.1103/PhysRevD.71.063004

Part II
Superluminality

Chapter 4
Superluminal Propagation in Galileon Models

Since the Galileon has been introduced, they witnessed a plethora of investigations. They exhibit a broad and interesting phenomenology. However there is also a potentially worrying phenomenon, namely the fluctuations of the Galileon field can propagate superluminally in the regime of interest (Nicolis et al. 2009; Goon et al. 2011), i.e. faster than light. This will be the main focus of this chapter. Since the Galileon interactions arise in a natural way in massive gravity, the theories of massive gravity share the same destiny. So superluminal modes are generic to Galileon and massive gravity constructions. The theory of special relativity does not accommodate propagations faster then the speed of light, therefore one often encounters concerns in the literature about theories yielding propagation faster than light. Superluminal fluctuations are considered to be a symptom of a sick theory. This is still an ongoing debate which we will not comment in detail about. Instead, our aim here is to prove how some class of modified gravities unavoidably give rise to superluminal propagation. Specially we will focus on the Galileon theories. The superluminal propagation in the single field Galileon models has been already shown in Nicolis et al. (2009). We will summarize the essence of the calculation and refer for more detail to the original paper (Nicolis et al. 2009). The single Galileon scalar field theory has been generalized to a multitude of interacting Galileon fields whose origin again can be traced back to Lovelock invariants in the higher co-dimension bulk, Hinterbichler et al. (2010), or such as in Cascading Gravity, de Rham et al. (2008), Padilla et al. (2010, 2011). Furthermore, they have been extended to arbitrary even p-forms whose field equations still only contain second derivatives (Deffayet et al. 2010). In the literature, it was claimed that the generalization of the single Galileon field to Bi-Galileon, consisting of two coupled scalar fields with the corresponding invariance under internal galileon- and shifting transformations of the two fields, avoid superluminal propagation (Padilla et al. 2010, 2011), which has attracted much attention. A priori there is no reason what so ever why the Bi-Galileon should avoid superluminal propagation. In this chapter we will scrutinize the superluminality in Multi-Galileon theories and argue that in these models, the existence of the Vainshtein mechanism about a static spherically symmetric source comes hand in hand with the existence of superluminal modes. We will show that propagation of superluminal fluctuations is a common and

L. Heisenberg, *Theoretical and Observational Consistency of Massive Gravity*,
Springer Theses, DOI 10.1007/978-3-319-18935-2_4

unavoidable feature in Bi- and Multi-Galileon theories, unlike previously claimed in the literature. We will start with the proof of superluminal propagation in the case of the Cubic Galileon for a localized point-source. We show that the mere presence of Cubic Galileon interactions guaranties the superluminal propagation of modes in either the radial or the orthoradial direction far away from a point source. This is intrinsic to the fact that for such configurations, at least one field falls as $1/r$ at large distance as expected from the Coulomb potential. For that behaviour, the matrix encoding the temporal perturbations at the order $1/r^3$ vanishes at infinity while the orthoradial and radial perturbations arise with opposite sign. This property is independent of the number of Galileon fields present. We then push the proof further for the case of the Quartic Galileon for extended static spherically symmetric sources. Since the previous result is ubiquitous to any Cubic Galileon interactions, the only possible way to avoid superluminalities at large distances, is to set all the Cubic Galileon interactions to zero. In that case, we show that the Quartic Galileon always lead to some superluminalities at large distances in either the radial or the orthoradial direction again when considering a sensible extended source. Even if we restrict ourselves to point sources, we show that the Quartic Bi-Galileon always lead to the propagation of at least one superluminal radial mode for some range of r. This result contradicts previous claims found in the litterature. We shall emphasize that our result relies crucially on the assumption that (1) the field decay as the Coulomb potential at infinity, (2) that no ghost are present and (3) that the Vainshtein mechanism is active (i.e. the Quartic Galileon interactions dominate over quadratic kinetic terms near the source), i.e. using the same assumptions as the ones used previously in the literature. The derivation of our generic result relies on the interplay between the behaviour of the field at large and at small distances. Finally, we show that near a localized source superluminalities are also present in the radial direction in a theory which includes only the Cubic Galileon. In the following we will first recapitulate the superluminal propagation in the single Galileon model before we move on to the Bi- and Multi-Galileon theories.

4.1 Single Galileon

Let us first review the equations of motion for the single field Galileon (1.86) in the spherical coordinates and assume a localized point source $\rho = M\delta^3(\vec{r})$. The equations of motion for the single Galileon π in the static spherically backgrounds becomes

$$\frac{1}{r^2}\frac{d}{dr}\left\{r^3\left(c_2\frac{\pi'}{r}+2c_3(\frac{\pi'}{r})^2+2c_4(\frac{\pi'}{r})^3\right)\right\}=\frac{M}{M_{Pl}}\delta^3(\vec{r}) \tag{4.1}$$

where π' is the derivative with respect to the radial coordinate r. (Note that for a time independent Galileon field the equation of motion for $\mathcal{L}_5$ is exactly zero). This equation is equivalent to

$$c_2 \frac{\pi'}{r} + 2c_3 (\frac{\pi'}{r})^2 + 2c_4 (\frac{\pi'}{r})^3 = \frac{M}{4\pi r^3 M_{Pl}} \tag{4.2}$$

In order to have a healthy spherical solution the coefficients have to fulfill the following conditions

$$\begin{aligned} c_2 &> 0 \\ c_4 &\geq 0 \\ c_3 &> -\sqrt{\frac{3}{2} c_2 c_4}. \end{aligned} \tag{4.3}$$

Now we have to study the stability of the perturbations around this radial solution $\pi_0(r)$. For that we expand the Lagrangian to quadratic order in perturbations $\pi = \pi_0(r) + \delta\pi(t, \vec{r})$

$$\mathcal{L} = \frac{1}{2} \partial_t \delta\pi \mathcal{K}^{tt} \partial_t \delta\pi - \frac{1}{2} \partial_r \delta\pi \mathcal{U}^{rr} \partial_r \delta\pi - \frac{1}{2} \partial_\Omega \delta\pi \mathcal{V}^{\Omega\Omega} \partial_\Omega \delta\pi \tag{4.4}$$

with the matrices given by

$$\begin{aligned} \mathcal{U}_{rr} &= c_2 + 4c_3 \frac{\pi_0'}{r} + 6c_4 (\frac{\pi_0'}{r})^2 \\ \mathcal{K}_{tt} &= \frac{c_2^2 + 4c_2c_3 \frac{\pi_0'}{r} + 12(c_3^2 - c_2c_4)(\frac{\pi_0'}{r})^2 + 24(c_3c_4 - 2c_2c_5)(\frac{\pi_0'}{r})^3}{c_2 + 4c_3 \frac{\pi_0'}{r} + 4c_4 (\frac{\pi_0'}{r})^2} \\ &+ \frac{12(3c_4^2 - 4c_3c_5)(\frac{\pi_0'}{r})^4}{c_2 + 4c_3 \frac{\pi_0'}{r} + 4c_4 (\frac{\pi_0'}{r})^2} \\ \mathcal{V}_{\Omega\Omega} &= \frac{c_2^2 + 2c_2c_3 \frac{\pi_0'}{r} + (4c_3^2 - 6c_2c_4)(\frac{\pi_0'}{r})^2}{c_2 + 4c_3 \frac{\pi_0'}{r} + 6c_4 (\frac{\pi_0'}{r})^2} \end{aligned} \tag{4.5}$$

The stability of the perturbations require $\mathcal{K}_{tt} > 0$ etc., which is only fulfilled if the parameters are constrained to

$$\begin{aligned} c_2 &> 0 \\ c_4 &\geq 0 \\ c_3 &\geq \sqrt{\frac{3}{2} c_2 c_4} \\ c_5 &\leq \frac{3}{4} \frac{c_4^2}{c_3} \end{aligned} \tag{4.6}$$

Taking into account these stability conditions we can now study the propagation speed of fluctuations. The radial speed of fluctuations is given by[1]

[1] Stability conditions enforce superluminal propagation of radial fluctuations while subluminal propagation in the angular direction.

$$c_r^2 = \frac{\mathcal{U}_{rr}}{\mathcal{K}_{tt}} = 1 + 4\frac{c_3}{c_2}\frac{\pi_0'}{r} + \cdots > 1 \tag{4.7}$$

which is always superluminal in order to fulfill the stability conditions while the angular speed of fluctuations is forced to be subluminal $c_\Omega^2 = \mathcal{V}_{\Omega\Omega}/\mathcal{K}_{tt} < 1$. For the single Galileon field the superluminal propagation is an unavoidable consequence for the fluctuations to be stable.

4.2 Multi-Galileon

In the following we will first review the formalism of the Bi- and Multi-Galileon and hereby adopt the same notation as in Padilla et al. (2010). We start first with the analysis needed for the study of the propagation of fluctuations around spherical symmetric backgrounds. We study the perturbations around the background generated by a point mass at large distances from that source. We show that there is always one mode which propagates superluminally whenever at least one Cubic Galileon interaction is present, regardless of the number of Galileons present in the theory. We also find that there are sensible source distributions around which there will always be a superluminal mode at large distances even if all the Cubic Galileon interactions are absent, for any number of Galileons. We then consider more closely the case of a point mass source when all the Cubic Galileon interactions are absent. In particular we study the short distance behaviour around a point mass background in the Bi-Galileon theory, and we find that there is again always a superluminal mode. In the case of vanishing asymptotic conditions $\pi \to 0$ the existence of the Vainshtein mechanism comes hand in hand with the existence of superluminal modes. This constitutes a No-go theorem showing that superluminal modes are generic to Galileon theories.

We consider the most general Multi-Galileon theory, in four dimensions. This model consists of N coupled scalar fields, $\pi_1, \ldots, \pi_N$. For simplicity we neglect gravity in our analysis and study the theory on Minkowski space-time. Similarly to Galileon theories (Nicolis et al. 2009), the Multi-Galileon theory is invariant under internal Galilean and shift transformations

$$\begin{aligned} &\pi_1 \to \pi_1 + b_1^\mu x_\mu + c_1 \\ &\ldots \\ &\ldots \\ &\pi_N \to \pi_N + b_N^\mu x^\mu + c_N \end{aligned}$$

If we consider the Galileon scalar fields as scalar fields in their own right, they could in principle couple to matter in a number of different ways. At the linear level, they can either couple as a conformal mode, $\pi_i T$ or as a longitudinal mode, $\partial_\mu \partial_\nu \pi_i T^{\mu\nu}$. However when dealing with conserved matter sources, this coupling vanishes and is thus irrelevant, no matter how many fields are considered. At the non-linear level, one can consider more general types of conformal couplings of the form $f(\pi_1, \dots, \pi_N)T$. Notice however that a general non-linear conformal coupling of that form breaks the Galileon symmetry at the level of the equations of motion directly. In other words, if one considers a general conformal coupling of the form $f(\pi_1, \dots, \pi_N)T$, one could always perform a field redefinition $\pi_i \to \hat{\pi}_i$ of the form $\hat{\pi}_1 = f(\pi_1, \dots, \pi_N)$ and $\hat{\pi}_j = \pi_j$ for $j = 2, \dots, N$, such that only $\hat{\pi}_1$ couples to matters in a linear way. The field redefinition $\hat{\pi}_1 = f(\pi_1, \dots, \pi_N)$ might imply that the field interactions are no longer of the Galileon type, which is just another way to see that an arbitrary non-linear coupling of the form $f(\pi_1, \dots, \pi_N)T$ is not part of the Galileon family. Finally non-linear couplings of the form $f_1(\pi_1, \dots, \pi_N)\partial_\mu \pi_i \partial_\nu \pi_j T^{\mu\nu}$ can also be considered. For instance the coupling of the $\partial_\mu \pi \partial_\nu \pi T^{\mu\nu}$ is generic in Massive Gravity (de Rham et al. 2010). However such a type of coupling cancels at the background level for static spherically symmetric sources, and are thus irrelevant to the present discussion. In conclusion, only the conformal coupling to external matter is important when dealing with conserved and static spherically symmetric sources and analyzing the behaviour of the perturbations in the vacuum. Whilst in principle the conformal coupling could be fully non-linear, only the linear one respects the Galileon symmetry at the level of the equations of motion. So in what follows we only focus on this linear conformal coupling, without any loss of generality. Furthermore, we couple only one of the N scalar fields to the trace of the stress energy tensor, as one can always rotate the field space $\pi_1, \dots, \pi_N$ to do so.

Considering the previous linear conformal coupling to matter, the most general multi-Galileon Lagrangian in four dimensions is

$$\mathcal{L} = \mathcal{L}_2 + \mathcal{L}_3 + \mathcal{L}_4 + \mathcal{L}_5 + \pi_1 T\,, \tag{4.8}$$

where the respective Quadratic $\mathcal{L}_2$, Cubic $\mathcal{L}_3$, Quartic $\mathcal{L}_4$ and Quintic Galileon $\mathcal{L}_5$ interactions are given by

$$\mathcal{L}_n(\pi_1, \dots, \pi_N) = \sum_{m_1+\dots+m_N=n-1} \mathcal{L}_{m_1,\dots,m_N} \tag{4.9}$$

with

$$\mathcal{L}_{m_1,\dots,m_N} = (\alpha^1_{m_1,\dots,m_N}\pi_1 + \dots + \alpha^N_{m_1,\dots,m_N}\pi_N)\mathcal{E}_{m_1,\dots,m_N}\,, \tag{4.10}$$

where the $\alpha^n_{m_1,\dots,m_N}$ are the coefficients for the Galileon interactions. Notice that this parameterization allow for a lot of redundancy, so not all the $\alpha^n_{m_1,\dots,m_N}$ are meaningful (many of them can be set to zero without loss of generality). Notice as well that in this language these coefficients α's are dimensionfull (the dimension depends on $m_1 + \dots + m_N$). We stick nonetheless to this notation for historical reasons, Padilla et al. (2010). In this formalism, all the derivative are included in the $\mathcal{E}_{m_1,\dots,m_N}$ which can be expressed as

$$\mathcal{E}_{m_1,\dots,m_N} = (m_1 + \cdots + m_N)!\delta^{\mu_1}_{[\nu_1} \cdots \delta^{\mu_{m_1}}_{\nu_{m_1}} \cdots \delta^{\rho_1}_{\sigma_1} \cdots \delta^{\rho_{m_N}}_{\sigma_{m_N}]} \times \left[(\partial_{\mu_1} \partial^{\nu_1} \pi_1) \cdots (\partial_{\mu_{m_1}} \partial^{\nu_{m_1}} \pi_1) \right] \cdots \times \left[(\partial_{\rho_1} \partial^{\sigma_1} \pi_N) \cdots (\partial_{\rho_{m_N}} \partial^{\sigma_{m_N}} \pi_N) \right], \tag{4.11}$$

using the formalism derived in Deffayet et al. (2009).

4.2.1 Bi-Galileon

Specializing this to the Bi-Galileon $N = 2$ is straightforward. The analogue to (4.10) for the Bi-Galileon would simply be

$$\mathcal{L}_{\pi_1,\pi_2} = \sum_{0 \le m+n \le 4} (\alpha_{m,n}\pi_1 + \beta_{m,n}\pi_2)\mathcal{E}_{m,n}, \tag{4.12}$$

with

$$\mathcal{E}_{m,n} = (m+n)!\delta^{\mu_1}_{[\nu_1} \cdots \delta^{\mu_m}_{\nu_m} \delta^{\rho_1}_{\sigma_1} \cdots \delta^{\rho_m}_{\sigma_n]} (\partial_{\mu_1} \partial^{\sigma_{\nu_n}} \pi_1) \times \cdots (\partial_{\mu_m} \partial^{\nu_m} \pi_1)(\partial_{\rho_1} \partial^{\sigma_1} \pi_2) \cdots (\partial_{\rho_n} \partial^{\sigma_n} \pi_2). \tag{4.13}$$

The equations of motion for the two scalar fields π_1 and π_2 are then

$$\sum_{0 \le m+n \le 4} a_{m,n}\mathcal{E}_{m,n} = -T \quad \text{and} \quad \sum_{0 \le m+n \le 4} b_{m,n}\mathcal{E}_{m,n} = 0, \tag{4.14}$$

where the coefficients $a_{m,n}$ and $b_{m,n}$ can be expressed in terms of the parameters $\alpha_{m,n}$ and $\beta_{m,n}$ as

$$a_{m,n} = (m+1)(\alpha_{m,n} + \beta_{m+1,n-1}) \quad \text{and} \quad b_{m,n} = (n+1)(\beta_{m,n} + \alpha_{m-1,n+1}). \tag{4.15}$$

We refer to Padilla et al. (2010, 2011) for more detailed discussions.

4.3 Spherical Symmetric Backgrounds

In this section, we recapitulate the formalism needed to study the superluminality of fluctuations about spherical symmetric solutions. For this we split every field into a spherically symmetric background configuration $\pi^0(r)$ and a fluctuation $\delta\pi(t, \vec{r})$,

$$\pi_n(t, \vec{r}) = \pi^0_n(r) + \delta\pi_n(t, \vec{r}), \qquad \forall\ n = 1, \dots, N, \tag{4.16}$$

and introduce the N-dimensional fluctuation vector in field space,

$$\Pi(t,\vec{r}) = \begin{pmatrix} \delta\pi_1(t,\vec{r}) \\ \vdots \\ \delta\pi_N(t,\vec{r}) \end{pmatrix} . \tag{4.17}$$

Similarly as in the single field Galileon case, at quadratic order in the fluctuations, the Lagrangian can be written as

$$\mathcal{L}_{\pi_1,\ldots,\pi_N} = \frac{1}{2}\partial_t\Pi\,.\,\mathcal{K}\,.\,\partial_t\Pi - \frac{1}{2}\partial_r\Pi\,.\,\mathcal{U}\,.\,\partial_r\Pi - \frac{1}{2}\partial_\Omega\Pi\,.\,\mathcal{V}\,.\,\partial_\Omega\Pi\,. \tag{4.18}$$

The kinetic matrix $\mathcal{K}$ and the two gradient matrices $\mathcal{U}$ and $\mathcal{V}$ are defined as follows:

$$\mathcal{K} = (1+\frac{r}{3}\partial_r)\,(\Sigma_1 + 3\Sigma_2 + 6\Sigma_3 + 6\Sigma_4) \tag{4.19}$$

$$\mathcal{U} = \Sigma_1 + 2\Sigma_2 + 2\Sigma_3 \tag{4.20}$$

$$\mathcal{V} = (1+\frac{r}{2}\partial_r)\mathcal{U}\,, \tag{4.21}$$

where the Σ matrices depend on the spherically symmetric background configuration (and are thus functions of r). In this language the nth matrix Σ_n encodes information about the $(n+1)$th order Galileon interactions $\mathcal{L}_{n+1}$,

$$\Sigma_n = \begin{pmatrix} \partial_{y_1} f_n^{a_1} & \cdots & \partial_{y_N} f_n^{a_N} \\ \vdots & \ddots & \vdots \\ \partial_{y_1} f_n^{a_1} & \cdots & \partial_{y_N} f_n^{a_N} \end{pmatrix} , \tag{4.22}$$

with

$$f_n^{\alpha}(y_1(r),\ldots,y_N(r)) = \sum_{i=0}^{n} (\alpha_{i,n-i}^{1'} + \alpha_{i,n-i}^{N'})y_1^i(r)\cdots y_N^{n-i}(r)\,, \tag{4.23}$$

and for each of the Galileon field, we define,

$$y_n(r) = \frac{\partial_r \pi_n^0(r)}{r} . \tag{4.24}$$

In terms of the matrix $\mathcal{U}$, the background equations of motion are given by

$$\frac{1}{r^2}\partial_r\left(r^2\mathcal{U}(\pi^0(r)).\partial_r \begin{pmatrix} \pi_1^0(r) \\ \pi_2^0(r) \\ \vdots \\ \pi_N^0(r) \end{pmatrix}\right) = -\begin{pmatrix} T \\ 0 \\ \vdots \\ 0 \end{pmatrix} . \tag{4.25}$$

In particular for a point source of mass $M = 4\pi m$ localized at the origin $r = 0$, we have

$$(\Sigma_1 + 2\Sigma_2(r) + 2\Sigma_3(r)) . \begin{pmatrix} y_1(r) \\ y_2(r) \\ \vdots \\ y_N(r) \end{pmatrix} = \begin{pmatrix} \frac{m}{r^3} \\ 0 \\ \vdots \\ 0 \end{pmatrix} , \tag{4.26}$$

where Σ_1 is independent of y_n and is thus simply a constant, Σ_2 is linear in the y_n and Σ_3 is quadratic in the fields.

Notice that the expressions (4.19–4.21) for the matrices $\mathcal{K}, \mathcal{U}$ and $\mathcal{V}$ in terms of the Σ_n matrices are universal and do not depend on the number N of fields.

4.3.1 *Focus on the Bi-Galileon*

In the following we restrict our attention to the Bi-Galileon since we will first focus on that case and then generalize our results to the Multi-Galileon case. In the Bi-Galileon case, the matrices Σ_n are given explicitly as below:

$$\Sigma_n = \begin{pmatrix} \partial_y f_n^a & \partial_y f_n^b \\ \partial_z f_n^a & \partial_z f_n^b \end{pmatrix} \qquad \text{with} \qquad f_n^\alpha = \sum_{i=0}^{n} \alpha'_{i,n-i} y^i z^{n-i} \tag{4.27}$$

The functions y and z appearing in the coefficients f_n^α are shortcuts for

$$y(r) = \frac{1}{r}\frac{\partial \pi_1^0}{\partial r} \tag{4.28}$$

$$z(r) = \frac{1}{r}\frac{\partial \pi_2^0}{\partial r} , \tag{4.29}$$

such that the equations of motion for π_1^0 and π_2^0 become simply

$$f_1^a + 2(f_2^a + f_3^a) = \frac{m}{r^3} \tag{4.30}$$

$$f_1^b + 2(f_2^b + f_3^b) = 0 , \tag{4.31}$$

where $m = M/4\pi$, and M is the mass of the point particle introduced at $r = 0$. More explicitly, in terms of y and z the two equations of motion are given by

$$a_{10}y + a_{01}z + 2\left(a_{20}y^2 + a_{11}yz + a_{02}z^2\right) + 2\left(+a_{30}y^3 + a_{21}y^2z + a_{12}yz^2 + a_{03}z^3\right) = \frac{m}{r^3} , \tag{4.32}$$

$$\mathrm{a}_{01}\mathrm{y} + \mathrm{b}_{01}\mathrm{z} + 2\left(\frac{\mathrm{a}_{11}}{2}\mathrm{y}^2 + 2\mathrm{a}_{02}\mathrm{yz} + \mathrm{b}_{02}\mathrm{z}^2\right)$$
$$+ 2\left(\mathrm{a}_{21}/3\mathrm{y}^3 + \mathrm{a}_{12}\mathrm{y}^2\mathrm{z} + 3\mathrm{a}_{03}\mathrm{yz}^2 + \mathrm{b}_{03}\mathrm{z}^3\right) = 0\,. \tag{4.33}$$

In terms of the parameters a_{ij} and b_{ij}, the $\Sigma_{1,2,3}$ matrices can then be expressed respectively as

$$\Sigma_1 = \begin{pmatrix} a_{10} & a_{01} \\ a_{01} & b_{01} \end{pmatrix}, \tag{4.34}$$

$$\Sigma_2 = \begin{pmatrix} 2a_{20}y + a_{11}z & a_{11}y + 2a_{02}z \\ a_{11}y + 2a_{02}z & 2a_{02}y + 2b_{02}z \end{pmatrix}, \tag{4.35}$$

$$\Sigma_3 = \begin{pmatrix} 3a_{30}y^2 + 2a_{21}yz + a_{12}z^2 & 3a_{03}z^2 + 2a_{12}zy + a_{21}y^2 \\ 3a_{03}z^2 + 2a_{12}zy + a_{21}y^2 & 3b_{03}z^2 + 6a_{03}zy + a_{12}y^2 \end{pmatrix}, \tag{4.36}$$

in the Bi-Galileon case. To get these expressions we have used the fact that for $m < n$ we have the correspondences $\mathcal{E}_{m,n} = \mathcal{E}_{n,m}|_{\pi_1^0 \leftrightarrow \pi_2^0}$. The exclusion of superluminal mode propagation implies that the sound speed of both modes along both the radial and orthoradial directions be less than or exactly equal to 1. The two sound speeds in the radial direction are given by the eigenvalues of the matrix $\mathcal{M}_r = \mathcal{K}^{-1}\mathcal{U}$ and the two sound speeds along the orthoradial direction are given by the eigenvalues of the matrix $\mathcal{M}_\Omega = \mathcal{K}^{-1}\mathcal{V}$. Therefore the condition for no superluminality is equivalent to requiring that all the eigenvalues of both matrices $\mathcal{M}_r - \mathbb{I}$ and $\mathcal{M}_\Omega - \mathbb{I}$ be zero or negative (and larger than -1), with $\mathbb{I}$ the identity matrix. In the following sections we study the behavior of the system in two different regimes, in the large and short distance regimes and confirm explicitly that there always exists at least one superluminal mode in one direction.

4.4 Superluminalities at Large Distances

This section is devoted to the behavior at large distances, specially we will study how the superluminal propagation appears through the existence of Cubic or Quartic Galileon at large distances. We will first show that if at least one Cubic Galileon interaction is present, then superluminal propagation is always present at large enough distances from a point source. It can be an interaction involving just one of the N Galileon fields, or an interaction mixing different Galileon fields together, the result remains unchanged. One way to bypass this conclusion might be to remove all the Cubic Galileon interactions for all N fields $\mathcal{L}_3 \equiv 0$, meaning that any $\alpha^n_{m_1,\ldots,m_N}$ with $n = 1, \ldots, N$ and $m_1 + \cdots + m_N = 2$ has to vanish exactly (for example in the Bi-Galileon case, this implies $a_{20} = a_{11} = a_{02} = a_{02} = 0$). If the coefficients are merely small, then one can always go to large enough distances where the Cubic Galileon dominates over the Quartic and Quintic Galileon interactions. Nevertheless even if all the Cubic Galileon terms vanish $\mathcal{L}_3 \equiv 0$, we can still find perfectly sensible static, spherically symmetric matter distributions around which there are superluminalities due to the Quartic Galileon at large distances. As a consequence, we will see in this section that as soon as either a Cubic or a Quartic Galileon interaction is present one can always construct a sensible matter distribution which forces at least one of the N Galileon fields to propagate superluminally in one direction (either the radial or the orthoradial one). We emphasize again that the Quintic Galileon interactions $\mathcal{L}_5$ always vanish at the background level around static, spherically symmetric sources, independently of the number of fields, so that if one tries to avoid the above conclusions by making both the Cubic and the Quartic Galileon vanish, then there is no Vainshtein mechanism at all about these configurations.

4.4.1 Superluminalities from the Cubic Galileon

In the Multi-Galileon case, the background equations of motion for a point source at $r = 0$ are given in (4.26). At this point it is worth to mention that we assume trivial asymptotic conditions at infinity which implies that the Galileon interactions ought to die out at large distances. The contributions from Σ_1 are thus the leading ones at large distances. Consistency of the theory requires that $\det \Sigma_1 \neq 0$ (so that the theory does indeed exhibit N degrees of freedom) and the matrix Σ_1 is thus invertible. Therefore at large distances, the background equations of motion simplify to

$$\Sigma_1 \cdot \begin{pmatrix} y_1(r) \\ y_2(r) \\ \vdots \\ y_N(r) \end{pmatrix} = \begin{pmatrix} \frac{m}{r^3} \\ 0 \\ \vdots \\ 0 \end{pmatrix} . \tag{4.37}$$

Recalling that Σ_1 is invertible and independent of the field (Σ_1 is a constant), this implies that to leading order at large distance about a point-source, the fields die off as r^{-1}

$$y(r) \sim r^{-3} + \mathcal{O}(r^{-6}) \qquad \Rightarrow \qquad \pi^0(r) \sim r^{-1} + \mathcal{O}(r^{-4}) , \tag{4.38}$$

for at least one of the N fields, as expected from the Newtonian inverse square law which should be valid at infinity. As a result, at large distances the Σ matrices behave as follows:

$$\Sigma_1 = \bar{\Sigma}_1 \tag{4.39}$$

$$\Sigma_2 = \frac{1}{r^3} \bar{\Sigma}_2 + \mathcal{O}\left(r^{-6}\right) \tag{4.40}$$

$$\Sigma_3 = \frac{1}{r^6} \bar{\Sigma}_3 + \mathcal{O}(r^{-9}) , \tag{4.41}$$

where the 'barred' matrices $\bar{\Sigma}_{1,2,3}$ are independent of r.

As a direct result of this scaling, it is trivial to see that at large distances, the kinetic and gradient matrices $\mathcal{K}, \mathcal{U}$ and $\mathcal{V}$ are given by

$$\mathcal{K} = \bar{\Sigma}_1 + 0 + \mathcal{O}\left(\frac{1}{r^6}\right), \tag{4.42}$$

$$\mathcal{U} = \bar{\Sigma}_1 + \frac{2}{r^3}\bar{\Sigma}_2 + \mathcal{O}\left(\frac{1}{r^6}\right) \tag{4.43}$$

$$\mathcal{V} = \bar{\Sigma}_1 - \frac{1}{r^3}\bar{\Sigma}_2 + \mathcal{O}\left(\frac{1}{r^6}\right). \tag{4.44}$$

It is apparent that the perturbations at the order $\frac{1}{r^3}$ in the matrix $\mathcal{K}$ vanish while in the matrices $\mathcal{U}$ and $\mathcal{V}$ they always come with the opposite sign, hence there is always a superluminal direction at infinity.[2] These results coincide with what is already known in the case of one Galileon (Nicolis et al. 2009). This is intrinsic to the $\frac{1}{r^3}$ behaviour at infinity and to the presence of the Cubic Galileon, and is independent of the number of fields.

One possible way to bypass this very general result is to require the matrix $\bar{\Sigma}_2$ to vanish entirely, which could be for instance achieved by imposing all the Cubic Galileon interactions to vanish. At large enough distances the Cubic Galileon would always dominate over the Quartic one (assuming trivial asymptotic conditions at infinity), so imposing a hierarchy between the Cubic Galileon interactions and the other ones is not sufficient to avoid superluminalities. Therefore, one way would be if all the Cubic interactions would be completely absent. In particular, even if some eigenvalues of $\bar{\Sigma}_2$ vanish, the previous result remains unchanged, as long as $\bar{\Sigma}_2$ has at least one non-vanishing eigenvalue which would imply that the associated eigenmode in field space has a superluminal direction (either a radial or an orthoradial one). But if all the eigenvalues of $\bar{\Sigma}_2$ vanish, then we can in principale evade the previous argument, which for example can be accomplished by demanding all the coefficients arising from the cubic Galileon interactions to vanish exactly, in other words if the next to leading interactions arise from the Quartic Galileon.

In the very special case where $\bar{\Sigma}_2$ vanishes entirely (i.e. all its eigenvalues are identically 0), then the previous argument needs special care. The contribution from $\bar{\Sigma}_3$ implies that $(\mathcal{K}^{-1}\mathcal{U})$ and $(\mathcal{K}^{-1}\mathcal{V})$ do not necessarily have opposite sign. In any Multi-Galileon theory one can always tune the coefficients of $\mathcal{L}_3$ so that the matrix $\bar{\Sigma}_2$ vanishes identically and so the r^{-3} scaling is not the leading order correction to $\mathcal{U}$ and $\mathcal{V}$. For example, for the Bi-Galileon if the parameters of the theory are carefully chosen so as to satisfy $a_{20} = b_{02}c^3$, $a_{11} = 2b_{02}c^2$ and $a_{02} = b_{02}c$ with $c = a_{01}/b_{01}$ then $\mathcal{U}$ and $\mathcal{V}$ vanish identically at $\mathcal{O}(r^{-3})$ for a pure point source and the argument given above breaks down. However as soon as we consider an extended source with energy density going as $1/r^{3-\epsilon}$ would revive $\bar{\Sigma}_2$ and the argument would then again

[2]The presence of the Cubic Galileon guaranties the presence of superluminal propagation at large distances from a point source.

be the one above. So even for these special coefficients, there is a whole class of otherwise physically sensible solutions which exhibit superluminal propagation.

In Sect. 4.5 we consider this case more closely in the Bi-Galileon scenario and find that there is still always at least a superluminal mode for some range of r. For instance superluminalities unavoidably arise near the origin through the Quartic Galileon, unlike what was claimed in Padilla et al. (2011). But first we point out that one can very easily construct an extended source for which superluminalities are present at large distance for the Quartic Galileons just in the same way as they were for the Cubic ones.

4.4.2 Superluminalities from the Quartic Galileon About an Extended Source

When the Cubic Galileon is absent, the presence of superluminalities about a point-source is more subtle to prove and will be done explicitly in the next section. Nevertheless, even if the coefficients in the Cubic Galileon vanish, we can always find a background configuration in which we can see superluminalities at large distances arise using the same argument as for the Cubic Galileon. In particular we can consider a gas of particles with a spatially varying density of the form

$$T = M_0 \left(\frac{r_0}{r}\right)^{3/2} \tag{4.45}$$

where r_0 characterizes the scale over which the density varies and M_0 controls the overall strength of the density profile. This matter distribution can always be imagined for some arbitrarily large radius before being cut off.

In this case the asymptotic behavior of the background fields becomes

$$y_n(r) = \frac{Y_n^{(1)}}{r^{3/2}} + \frac{Y_n^{(2)}}{r^{9/2}} + \mathcal{O}(\frac{1}{r^{15/2}}), \tag{4.46}$$

for all the fields $n = 1, \ldots, N$, and we find once again using Eq. (4.19) that to order $\mathcal{O}(r^{-3/2})$, $\mathcal{K}$ vanishes and $\mathcal{U}$ and $\mathcal{V}$ have opposite signs, guaranteeing a superluminal direction.[3]

This illustrates the basic reason we expect any theory that exhibits the Vainshtein mechanism to inevitably contain superluminalities when considering trivial asymptotic conditions. Every new source configuration gives rise to a new background Galileon field configuration. Because the theory must be nonlinear in order to have a Vainshtein mechanism, the fluctuations around this background will propagate on aneffective metric determined by the background. Since the sources are not

[3]The presence of the Quartic Galileon guaranties the presence of superluminal propagation at large distances from an extended source.

constrained by the theory, we are free to choose any source we like, and so we have a lot of freedom to change the parameters in this background metric and create superluminalities.

4.5 Quartic Galileon About a Point-Source

In the previous section we have shown that superluminalities are ubiquitous in Galileon models. No matter the number of field there is no possible choice of parameters that can ever free the theory from superluminal propagation. The argument in the previous section was completely generic and independent of the number of fields. It only required the behaviour at large distances (when the non-linear Galileon interactions can be treated perturbatively).

In what follows we show that even for a point source in the Quartic Bi-Galileon model, superluminalities can never be avoided in a consistent model. Again, the only requirements we pose are the absence of ghost, the presence of an active Vainshtein mechanism, and trivial conditions at infinity. This result is in contradiction with previous results and examples in the literature, but upon presentation of this following argument, the previous claims have been reconsidered.

The philosophy of the argument goes as follows: We analyze the model both at large distances in the weak field limit and at short distances from the point source where the quartic interactions dominate (as required by the existence of an active Vainshtein mechanism). The requirement for stability (in particular the absence of ghost) sets some conditions on the parameters of the theory. We then show that these conditions are sufficient to imply the presence of superluminal modes near the source. We emphasize that this result could not be obtained, should we just have focused on the near origin behaviour without knowledge of the field stability at infinity.

4.5.1 Stability at Large Distances

To ensure the stability of the fields, the kinetic matrix $\mathcal{K}$ as well as the gradient matrices $\mathcal{U}$ and $\mathcal{V}$ should be positive definite at any point r. At infinity in particular these three conditions are equivalent and simply imply that the matrix Σ_1 ought to be positive definite. In the case of the Bi-Galileon, this implies

$$\det \Sigma_1 = \mathrm{a}_{10}\mathrm{b}_{01} - \mathrm{a}_{01}^2 > 0 \quad \text{and} \quad \mathrm{Tr}\, \Sigma_1 = \mathrm{a}_{10} + \mathrm{b}_{01} > 0\,. \tag{4.47}$$

In terms of the coefficients of the quadratic terms, these two conditions imply

$$\mathrm{a}_{10} > 0 \quad \text{and} \quad \mathrm{b}_{01} > \frac{\mathrm{a}_{01}^2}{\mathrm{a}_{10}} > 0\,. \tag{4.48}$$

The behaviour of the fields at large distance from a point source localized at $r = 0$ is determined by the coefficients of the quadratic terms, (or equivalently by Σ_1),

$$y(r) = \frac{Y_1}{r^3} + \mathcal{O}\left(\frac{1}{r^6}\right), \qquad \text{with} \qquad Y_1 = \frac{b_{01}}{\det\Sigma_1} m \tag{4.49}$$

$$z(r) = \frac{Z_1}{r^3} + \mathcal{O}\left(\frac{1}{r^6}\right), \qquad \text{with} \qquad Z_1 = \frac{-a_{01}}{\det\Sigma_1} m, \tag{4.50}$$

and as expected, we recover a Newtonian inverse square law behavior for each mode at infinity, namely $\partial_r\pi_1^0 \sim \partial_r\pi_2^0 \sim r^{-2}$. At this stage it is worth to mention that the stability condition (4.48) implies that $Y_1 > 0$, which is consistent with the fact that the force mediated by the one field π_1 that couples to matter is attractive.

4.5.2 Short Distance Behavior

We now study the field fluctuations at small distances near the source (i.e. at leading order in r, assuming we are well within the Vainshtein region). From the equations of motion (4.32), (4.33) after setting the coefficients of the cubic Galileon to zero near the origin, we infer the following expansion

$$y(r) = \frac{y_1}{r} + y_2 r + \mathcal{O}(r^3) \tag{4.51}$$

$$z(r) = \frac{z_1}{r} + z_2 r + \mathcal{O}(r^3), \tag{4.52}$$

with

$$a_{30}y_1^3 + a_{21}y_1^2 z_1 + a_{12}y_1 z_1^2 + a_{03}z_1^3 = \frac{m}{2} \tag{4.53}$$

$$\frac{a_{21}}{3}y_1^3 + a_{12}y_1^2 z_1 + 3a_{03}y_1 z_1^2 + b_{03}z_1^3 = 0. \tag{4.54}$$

Note that the $\mathcal{O}(r^0)$ terms vanish since the Cubic Galileon is not present.

Expanding Σ_3 in powers of r, we have

$$\Sigma_3 = \Sigma_3^{(l)} + \Sigma_3^{(nl)} + \cdots = \frac{1}{r^2}\tilde{\Sigma}_3^{(l)} + \tilde{\Sigma}_3^{(nl)} + \mathcal{O}\left(r^2\right), \tag{4.55}$$

where the leading order contribution to Σ_3 is given by

$$\tilde{\Sigma}_3^{(l)} = \begin{pmatrix} 3a_{30}y_1^2 + 2a_{21}y_1 z_1 + a_{12}z_1^2 & 3a_{03}z_1^2 + 2a_{12}y_1 z_1 + a_{21}y_1^2 \\ 3a_{03}z_1^2 + 2a_{12}y_1 z_1 + a_{21}y_1^2 & 3b_{03}z_1^2 + 6a_{03}y_1 z_1 + a_{12}y_1^2 \end{pmatrix}. \tag{4.56}$$

Solving the equation of motion (4.54) for b_{03} gives

$$b_{03} = (-a_{21}y^3 - 3a_{12}y^2z - 9a_{03}yz^2)/(3z^3). \tag{4.57}$$

Similarly solving the equation of motion (4.53) for a_{30} yields

$$a_{30} = ((m - 8a_{21}\pi r^3y^2z - 8a_{12}\pi r^3yz^2 - 8a_{03}\pi r^3z^3)/(8\pi r^3y^3)). \tag{4.58}$$

Using these expressions for b_{03}, a_{30} and introducing the combination $\mathcal{B}$ defined as follows:

$$\mathcal{B} = 3a_{03}z_1^2 + 2a_{12}y_1z_1 + a_{21}y_1^2, \tag{4.59}$$

we can then write $\tilde{\Sigma}_3^{(1)}$ simply as:

$$\tilde{\Sigma}_3^{(1)} = \begin{pmatrix} -\frac{z_1}{y_1}\mathcal{B} + \frac{3m}{2y_1} & \mathcal{B} \\ \mathcal{B} & -\frac{y_1\mathcal{B}}{z_1} \end{pmatrix}. \tag{4.60}$$

4.5.3 Stability at Short Distances

As mentioned previously, me must ensure that the eigenvalues of $\mathcal{K}$ are strictly positive. At small distances near the source, the matrix $\mathcal{K}$ can be expressed as

$$\mathcal{K} = \frac{2}{r^2}\tilde{\Sigma}_3^{(1)} + \mathcal{O}(1). \tag{4.61}$$

In terms of y_1, z_1 and $\mathcal{B}$, the absence of ghost near the origin implies the following conditions on the parameters

$$\det\tilde{\Sigma}_3 = -\frac{3}{2}\frac{\mathcal{B}m}{z_1} > 0 \tag{4.62}$$

$$\mathrm{Tr}\tilde{\Sigma}_3 = \frac{3mz_1 - 2\mathcal{B}(y_1^2 + z_1^2)}{y_1z_1} > 0, \tag{4.63}$$

which are equivalent to

$$y_1 > 0, \quad \text{and} \quad \frac{\mathcal{B}}{z_1} < 0. \tag{4.64}$$

This seems a priori in contradiction with the results presented in Padilla et al. (2011), where a specific example was provided. Analyzing this example in Appendix A.1 we find the presence of superluminalities in agreement with the argument presented in this section. We see in particular in that example that the modes are superluminal at large distances already. Furthermore in the example presented in

Padilla et al. (2011) there is a mode with the wrong kinetic term near the origin. Since at infinity, the eigenvalues of $\mathcal{K}$ are positive in this example, the kinetic term of one of the modes must vanish at a given r, which would imply strong coupling issues. No healthy theory can accept such a behavior within its own regime of validity, it therefore seems unlikely that the explicit coefficients chosen in Padilla et al. (2011) lead to a physically healthy model. We now use the stability conditions derived at both large and small distances to deduce the behaviour of the radial sound speed in this model.

4.5.4 Sound Speed Near the Source

Similarly as we did at large distances, we can now compute the 'radial sound speed' matrix $\mathcal{M}_r = \mathcal{K}^{-1}\mathcal{U}$ near the origin,

$$\mathcal{M}_r = \mathbb{I} - 2r^2(\tilde{\Sigma}_3^{(l)})^{-1}\tilde{\Sigma}_3^{(nl)} + \mathcal{O}(r^4)\,. \tag{4.65}$$

We note that unlike the Cubic Galileon case described in more detail below, the leading order behaviour of $\mathcal{M}$ is not manifestly superluminal. However, this is not enough to guarantee the absence of superluminal modes, we must carefully check the sign of the small $\mathcal{O}(r^2)$ correction term before making any conclusions. A simple formulation for the matrix $\Sigma_3^{(l)}$ is given in (4.60), and a similar expression for $\Sigma_3^{(nl)}$ can be found in an analogous way,

$$\tilde{\Sigma}_3^{(nl)} = \begin{pmatrix} -a_{10} - a_{01}\frac{z_1}{y_1} - 2\zeta\frac{z_1}{y_1} & 2\zeta \\ 2\zeta & -b_{01} - a_{01}\frac{y_1}{z_1} - 2\zeta\frac{y_1}{z_1} \end{pmatrix}, \tag{4.66}$$

with the notation:

$$\zeta = a_{21}y_1y_2 + a_{12}y_2z_1 + a_{12}y_1z_2 + 3a_{03}z_1z_2\,. \tag{4.67}$$

This allows us to compute the radial sound speed

$$c_{s\pm}^2 = 1 + r^2(a' \pm \sqrt{b'}) + \mathcal{O}(r^4)\,, \tag{4.68}$$

with a' and b' some coefficients that depend on y_1, z_1, $\mathcal{B}$, m and $(a, b)_{ij}$. So for both modes to be subluminal along the radial direction, the following conditions should be satisfied:

$$a' < 0, \qquad b' > 0 \quad \text{and} \quad a'^2 - b' > 0\,. \tag{4.69}$$

However as we shall see, these are not consistent with the stability conditions established previously.

The explicit form of the coefficients a$'$ and b$'$[4] is given by:

$$a' = \frac{-1}{3m\mathcal{B}y_1}\Big(3(a_{01}y_1 + 2\zeta y_1 + b_{01}z_1) - 2\mathcal{B}\mathcal{D}\Big) \tag{4.70}$$

$$a'^2 - b' = \frac{-8}{3m\mathcal{B}y_1}\Big((a_{10}y_1 + a_{01}z_1)(a_{01}y_1 + b_{01}z_1) + 2\zeta\mathcal{D}\Big), \tag{4.71}$$

with the notation

$$\mathcal{D} = a_{10}y_1^2 + 2a_{01}z_1y_1 + b_{01}z_1^2 . \tag{4.72}$$

We may re-express the quantity $\mathcal{D}$ as follows

$$\mathcal{D} = a_{10}y_1^2 + 2a_{01}z_1y_1 + b_{01}z_1^2 \tag{4.73}$$

$$= a_{10}\left(y_1 + \frac{a_{01}}{a_{10}}z_1\right)^2 + \frac{z_1^2}{a_{10}}\left(a_{10}b_{01} - a_{01}^2\right) > 0 . \tag{4.74}$$

Recall from the expression of the kinetic matrix $\mathcal{K}$ at infinity, that the following two conditions should be satisfied, (4.47): $a_{10} > 0$ and $\left(a_{10}b_{01} - a_{01}^2\right) > 0$, which directly implies that $\mathcal{D}$ is strictly positive. Knowing this, we check whether or not $a' < 0$ and $a'^2 - b' > 0$ which if true, would imply that both modes are sub-luminal.

We start with the requirement that $a' < 0$. This implies that $3m(a_{01}y_1 + 2\zeta y_1 + b_{01}z_1) - 2\mathcal{B}(a_{10}y_1^2 + 2a_{01}z_1y_1 + b_{01}z_1^2)$ has the same sign as $\mathcal{B}$. Once this condition is satisfied, we check the sign of $a'^2 - b' > 0$. This quantity is positive only if $\mathcal{F}$ has the opposite sign as $\mathcal{B}$, where

$$\mathcal{F} = (a_{10}y_1 + a_{01}z_1)(a_{01}y_1 + b_{01}z_1) + 2\zeta\mathcal{D} . \tag{4.75}$$

In what follows, we will start by assuming that z_1 is positive and show that in that case the condition to avoid any super-luminal modes cannot be satisfied. The same remains true if z_1 is assumed to be negative. We can therefore conclude that the Quartic Bi-Galileon interactions always produce a superluminal mode already in the configuration about a point source.

We recall from Eq. (4.64) that if $z_1 > 0$, the absence of ghost-like modes near the origin imposes the condition $\mathcal{B} < 0$. Furthermore, knowing that $\mathcal{D} = a_{10}y_1^2 + 2a_{01}z_1y_1 + b_{01}z_1^2 > 0$, we can infer that a$'$ negative only if

$$a_{01}y_1 + 2\zeta y_1 + b_{01}z_1 < \frac{2\mathcal{B}\mathcal{D}}{3m} < 0 . \tag{4.76}$$

Then using the fact that $\mathcal{D} = a_{10}y_1^2 + 2a_{01}z_1y_1 + b_{01}z_1^2 > 0$, this implies (knowing from (4.64) that $y_1 > 0$):

[4] It easy to check that b$'$ is always positive.

$$a_{01}y_1 + 2\zeta y_1 + b_{01}z_1 < 0 \quad \Rightarrow \quad \zeta < -\frac{1}{2}(a_{01} + b_{01}\frac{z_1}{y_1}) . \tag{4.77}$$

Finally to avoid any superluminal mode, the quantity $a'^2 - b'$ should also be positive. Since in this case $\mathcal{B}$ is negative, $a'^2 - b'$ has the same sign as $\mathcal{F}$, where

$$\begin{aligned}\mathcal{F} &= (a_{10}y_1 + a_{01}z_1)(a_{01}y_1 + b_{01}z_1) + 2\zeta\mathcal{D} \\ &< (a_{10}y_1 + a_{01}z_1)(a_{01}y_1 + b_{01}z_1) - (a_{01} + b_{01}\frac{z_1}{y_1})\mathcal{D} \\ &< -\frac{z_1}{y_1}(a_{01}y_1 + b_{01}z_1)^2 .\end{aligned} \tag{4.78}$$

Since $y_1 > 0$ and $z_1 > 0$ this implies that $\mathcal{F} < 0$. Since $a'^2 - b'$ has the same sign as $\mathcal{F}$, we can therefore conclude that if we assume z_1 to be positive and $a' < 0$, the quantity $(a'^2 - b')$ will also be negative, or in other words, there is one superluminal mode.[5] This argument was made assuming $z_1 > 0$, however it is straightforward to reproduce the same argument for negative z_1. If we choose for instance negative z_1 ($z_1 < 0$) then the condition coming from the absence of ghost-like instabilities Eq. (4.64) will require this time the opposite sign for $\mathcal{B}$, namely $\mathcal{B} > 0$ and therefore $\mathcal{F}$ will be a positive number $\mathcal{F} < -\frac{z_1}{y_1}(a_{01}y_1 + b_{01}z_1)^2$ while the expression $(a'^2 - b')$ in Eq. (4.71) will have the opposite sign to $\mathcal{F}$ and therefore again there would not be any choice of coefficients $(a, b)_{ij}$ to make both modes (sub)luminal. With this we have proven that there is no generic choice for the parameters a_{ij} and b_{ij} near the origin which would prevent the propagation of superluminal modes.

4.5.5 *Special Case of Dominant First Order Corrections*

In the previous section we proved that close to the source there is always one mode which propagates superluminaly in a generic theory where only the Quartic Galileon is present. However we made a technical assumption in Eq. (4.65) that $\tilde{\Sigma}_3^{(1)}$ was invertible, or equivalently that we did not make the special choice of parameters a_{ij}, b_{0i} which gives $\mathcal{B} = 0$ (implying that the leading order pieces in Σ_3 were strictly larger than the first order corrections). However we could consider a specific choice for which some of the leading order pieces of Σ_3 vanish and the subleading pieces become dominant. Therefore in this section we will examine this possible loophole more closely. We will find that even in this case one always finds that a superluminal mode is present. When $\mathcal{B} = 0$, $\tilde{\Sigma}_3^{(1)}$ takes the following trivial form:

$$\tilde{\Sigma}_3^{(1)} = \begin{pmatrix} \frac{3m}{2y_1} & 0 \\ 0 & 0 \end{pmatrix} . \tag{4.79}$$

[5] The presence of the Quartic Galileon guaranties the superluminal propagation at short distances from a point source.

The stability of the theory now depends not only on the leading behavior of the kinetic $\mathcal{K}$ and radial derivative $\mathcal{U}$ matrices, but also on the subleading behavior. In terms of the Σ matrices, $\mathcal{K}$ and $\mathcal{U}$ take the following form

$$\mathcal{K} = 2\frac{\tilde{\Sigma}_3^{(1)}}{r^2} + \Sigma_1 + 6\tilde{\Sigma}_3^{(nl)}, \qquad \mathcal{U} = 2\frac{\tilde{\Sigma}_3^{(1)}}{r^2} + \Sigma_1 + 2\tilde{\Sigma}_3^{(nl)}. \tag{4.80}$$

The theory is stable only if $\mathcal{K}$ and $\mathcal{U}$ have positive eigenvalues, or in other words only if the following three quantities are positive:

$$\begin{aligned} y_1 &> 0, \\ \lambda_1^2 \equiv -\frac{3m}{y_1 z_1^2}\left(6(a_{01}+2\zeta)y_1 z_1 + 5b_{01}z_1^2\right) &> 0, \\ \lambda_2^2 \equiv -\frac{3m}{y_1 z_1^2}\left(2(a_{01}+2\zeta)y_1 z_1 + b_{01}z_1^2\right) &> 0. \end{aligned} \tag{4.81}$$

Now we construct again the radial speed of sound matrix $\mathcal{M}_r \equiv \mathcal{K}^{-1}\mathcal{U}$ in this specific case with $\mathcal{B} = 0$. We can write the trace and determinant as

$$\begin{aligned} \mathrm{tr}\mathcal{M}_r &= \left(1 + \frac{\lambda_2^2}{\lambda_1^2}\right) + r^2\,\tau + \mathcal{O}(r^4), \\ \det\mathcal{M}_r &= \frac{\lambda_2^2}{\lambda_1^2} + r^2\delta + \mathcal{O}(r^4). \end{aligned} \tag{4.82}$$

where τ and δ are functions of the given parameters (however we will only need $\tau - \delta$ as shown below). The speed of sound is given by, to $\mathcal{O}(r^2)$,

- If $\lambda_1^2 > \lambda_2^2$

$$c_\pm^2 = \begin{cases} 1 + r^2\frac{\lambda_1^2}{\lambda_1^2-\lambda_2^2}(\tau-\delta), \\ \frac{\lambda_2^2}{\lambda_1^2} - r^2\frac{\lambda_1^2}{\lambda_1^2-\lambda_2^2}(\tau-\delta). \end{cases} \tag{4.83}$$

In this case we will have superluminal propagation if and only if $\tau - \delta > 0$. We show that one always has $\tau - \delta > 0$ in this case by carefully making use of the stability constraints in Appendix A.2.

- If $\lambda_1^2 < \lambda_2^2$

$$c_\pm^2 = \begin{cases} \frac{\lambda_2^2}{\lambda_1^2} + r^2\frac{\lambda_1^2}{\lambda_2^2-\lambda_1^2}(\tau-\delta), \\ 1 - r^2\frac{\lambda_1^2}{\lambda_2^2-\lambda_1^2}(\tau-\delta). \end{cases} \tag{4.84}$$

Since $\lambda_1^2 < \lambda_2^2$, in this case superluminal propagation is guaranteed.

One might argue that this only guarantees superluminalitiy at the origin which is inside the redressed strong coupling radius of the theory, where we can no longer trust the results of the theory. However, we note that explicitly factoring out powers of M and Λ that the speed of sound is given by

$$c_{\pm}^2 = \frac{\lambda_2^2}{\lambda_1^2} + \left(\frac{r}{r_V}\right)^2 \frac{\lambda_1^2}{\lambda_2^2 - \lambda_1^2}(\hat{\tau} - \hat{\delta}), \tag{4.85}$$

where $r_V \equiv (M/M_{Pl})^{1/3}/\Lambda$ is the Vainshtein radius and where $\hat{\tau}$ and $\hat{\delta}$ are dimensionless. Since $r_V > \Lambda^{-1}$, and since the redressed strong coupling radius is always smaller than Λ^{-1}, there is a range of r in which we can trust the theory and we can also trust the leading order behavior of the speed of sound above.

4.6 Cubic Lagrangian Near the Source

Lets have a quick look into the contributions coming from a Cubic Bi-Galileon theory near the origin and study the superluminality. In the Sect. 4.4.1 we had seen that the existence of the Cubic Galileon guarantees superluminal propagation at infinity. Now lets also see the effect of a pure Cubic Galileon term on short distances. We quickly recall the equations of motion in the Cubic Galileon case near the origin here again:

$$2\left(a_{02}z^2 + a_{11}yz + a_{20}y^2\right) = \frac{m}{r^3} \tag{4.86}$$

$$2\left(b_{02}z^2 + 2a_{02}yz + \frac{1}{2}a_{11}y^2\right) = 0\,. \tag{4.87}$$

At short distance the fields then behave as

$$y(r) = \frac{y_1}{r^{3/2}} + y_2 + \mathcal{O}(r^{3/2})\,, \tag{4.88}$$

$$z(r) = \frac{z_1}{r^{3/2}} + z_2 + \mathcal{O}(r^{3/2})\,. \tag{4.89}$$

The leading order matrix $\tilde{\Sigma}_2^{(1)}$ can be expressed as follows (after use of the equations of motion):

$$\tilde{\Sigma}_2^{(1)} = \begin{pmatrix} \frac{m}{y_1} - \frac{z_1}{y_1}\mathcal{C} & \mathcal{C} \\ \mathcal{C} & -\frac{y_1}{z_1}\mathcal{C} \end{pmatrix}. \tag{4.90}$$

with the notation

$$\mathcal{C} = a_{11}y_1 + 2a_{02}z_1\,. \tag{4.91}$$

The stability condition in the short distance regime requires $\det\mathcal{K} \approx \det\Sigma_2^{(l)} > 0$ and $\mathrm{Tr}\mathcal{K} \approx \mathrm{Tr}\tilde{\Sigma}_2^{(l)} > 0$:

$$\mathrm{Tr}\Sigma_2^{(l)} = \frac{m - \mathcal{C} z_1}{y_1} > 0 \tag{4.92}$$

$$\det\Sigma_2^{(l)} = \frac{-m\mathcal{C}}{z_1} > 0\,. \tag{4.93}$$

These conditions imply $\frac{\mathcal{C}}{z_1} < 0$ and $y_1 > 0$. After using the next-to-leading order equations of motion to simplify the next-to-leading order matrix $\tilde{\Sigma}_2^{(nl)}$, we find

$$\tilde{\Sigma}_2^{(nl)} = \begin{pmatrix} -\frac{a_{10}}{2} - a_{01}\frac{z_1}{2y_1} - 2\beta\frac{z_1}{y_1} & 2\beta \\ 2\beta & -\frac{b_{01}}{2} - \frac{a_{01}y_1}{2z_1} - 2\beta\frac{y_1}{z_1} \end{pmatrix}, \tag{4.94}$$

with $\beta \equiv a_{02}z_2 + \frac{a_{11}}{2}y_2$.

Assuming $\Sigma_2^{(l)}$ is invertible (which is the case if there is at least one non-vanishing Cubic Galieon interaction), the matrix $\mathcal{M}_r$ is given by

$$\begin{aligned}\mathcal{M}_r = \mathcal{K}^{-1}\mathcal{U} &= \left[\frac{3}{2}\frac{\Sigma_2^{(l)}}{r^{3/2}} + (\Sigma_1 + 3\Sigma_2^{(nl)})\right]^{-1}\left[2\frac{\Sigma_2^{(l)}}{r^{3/2}} + (\Sigma_1 + 2\Sigma_2^{(nl)})\right] \\ &= \frac{4}{3} - \frac{2}{3}r^{3/2}\left[\Sigma_2^{(l)}\right]^{-1}\left[\Sigma_1 + 4\Sigma_2^{(nl)}\right].\end{aligned} \tag{4.95}$$

This in turn implies that the Cubic Galileon also gives rise to superluminal propagation near the origin. If on the other hand we consider the possible loophole with vanishing determinant of the leading matrix $\Sigma_2^{(l)}$ (choosing parameters such $\mathcal{C} = 0$) nothing changes. The matrix $\mathcal{M}_r$ has still one eigenvalue going as $4/3 + \mathcal{O}(r)$, and another eigenvalue whose leading behavior depends on the signs and relative strengths of β and z_1. But the existence of one eigenvalue that is $4/3$ at leading order is enough to prove the existence of superluminalities in that regime as well.

4.7 Summary and Discussion

We have shown that Multi-Galileon theories inevitably contain superluminal modes around some backgrounds, for any number of Galileon fields. At large distances from a static point source, we have shown that if the Cubic Galileon is present (even if its coefficients are very small), it will eventually dominate over the other Galileons and lead to a superluminal mode. Even if no Cubic Galileon interactions are present (i.e. all the Cubic Galileon coefficients are exactly zero), we find that there are simple, perfectly valid matter distributions (such as a static gas of particles whose density falls of as $r^{-3/2}$) around which perturbations propagate superluminally.

We also considered the case studied in the journal version of Padilla et al. (2011) of perturbations around a static point source in the Bi-Galileon when only the Quartic Galileon is present. By studying the speed of sound of perturbations close to the source, we find, in contradiction to their original claims, that the presence of a superluminal mode cannot be avoided. This is a nontrivial result, which can only be seen by carefully taking into account the constraints that stability at large distances places on the theory, and the interplay between these conditions coming from infinity and the action for perturbations near the source. In other words, this is not a local result which could have been derived from the knowledge of the behaviour near the source only. We have also showed that there will always be superluminal perturbations around a point source if only a Cubic Galileon is present.

Our results physically arise from the link between the Vainshtein mechanism and superluminalities in typical Galileon theories. So long as one is considering theories that are ghost-free, with trivial asymptotic conditions at infinity and avoid quantum strong-coupling issues (fields with vanishing kinetic terms), these two effects are intimately connected. One way to see this link is to note that the Vainshtein mechanism is inherently nonlinear, and so the behavior of the perturbations depends strongly on the source distribution present. Thus one expects to always be able to find backgrounds around which there are superluminalities. However, the connection may be stronger than this: As we have shown, even in the case of a static point source with only a Quartic Galileon present, where the presence of superluminalities at large distances is not manifest, there are still inevitably superluminalities close to the source.

We believe that the superluminalities are a crucial feature of Galileon theories. As shown in Adams et al. (2006), the presence of superluminalities around some backgrounds is ultimately tied with the failure of the Galileon theory to have a Wilsonian completion. It would be interesting to understand whether this aspect and the presence of a Vainshtein mechanism could however be tied to theories which allow for an alternative to UV completion such as classicalization, Dvali et al. (2012, 2011), Vikman (2013).

Recently, a dual description to the Galileon interactions has been discovered (de Rham et al. 2013). This duality map the original Galileon theory to another Galileon theory by a non trivial field redefinition $\tilde{x}^\mu = x^\mu + \partial^\mu \pi(x)$. The inverse transformation of it defines the dual Galileon $x^\mu = \tilde{x}^\mu + \tilde{\partial}^\mu \rho(\tilde{x})$. For a given very specific Galileon coefficients it has been shown that the Galileon interactions are dual to a free massless scalar field. This mapping between a free luminal theory and the superluminal Galileon theory suggest that the naive existence of superluminal propagation can still give rise to causal theory with analytic and unitary S-matrix (de Rham et al. 2013). This duality maps a strongly coupled state to a weakly coupled state, therefore this technique can be used to perform the UV completion via classicalization.

The worry about the superluminal propagation comes from the fact that superluminal fluctuations could allow for acausality and build configurations with Close-Timelike-Curves (CTCs). Nevertheless there are cases in which the superluminal fluctuations come with their own metric and causal structure, which can be very different to that felt by photons, and the causal cones of these fluctuations might even lie outside the causal cones of photons. Regardless of all this, the causal structure of the spacetime can be protected (Babichev et al. 2008) if there exists one foliation of spacetime into surfaces which can be considered as Cauchy surfaces for both metrics. In theories of Galilean invariant interactions it is possible to construct CTCs within the naive regime of validity of the effective field theory (as is also the case in GR). Nevertheless, as we have shown in Burrage et al. (2012), the CTCs never arise since the Galileon inevitably becomes infinitely strongly coupled implying an infinite amount of backreaction. The backreaction on the background for the Galileon field breaks down the effective field theory and forbids the formation of the CTC through the backreaction on the spacetime geometry. The setup of background solutions with CTCs becomes unstable with an arbitrarily fast decay time. As a result, theories of Galilean invariant interactions satisfy a direct analogue of Hawking's chronology protection conjecture (Burrage et al. 2012).

We conclude by reviewing the only known way (so far) to have a Vainshtein mechanism and still avoid superluminalities. If the Galileon is not considered as a field in its own right, but rather as a component of anther fully fledge theory, one needs not to impose trivial asymptotic conditions at infinity. In massive gravity for instance, the Galileon field that appears in its decoupling limit is not a fundamental field. In such setups, it is then consistent to consider configurations for which the Galileon field does not vanish at infinity, so long as the metric is well defined at infinity (which does not necessarily imply Minkowski space-time). In such cases, we can thus have more freedom to fix the asymptotic boundary conditions for the Galileon field. A specific realization has recently been found in Berezhiani et al. (2013), where the asymptotic behaviour is non trivial and the metric asymptotes to a cosmological one at large distances. These results do rely on the existence of a non-trivial coupling to matter of the form $\partial_\mu \pi \partial_\nu \pi T^{\mu\nu}$ which naturally arises in Massive Gravity, de Rham et al. (2010). When such non-trivial asymptotics conditions are considered, the results derived in this work are no longer valid and open the door for a way to find configurations which do exhibit the Vainshtein mechanism without necessarily propagating a superluminal mode around these configurations. Future work should consider the role of boundary conditions in the selection of viable configurations.

References

Adams A, Arkani-Hamed N, Dubovsky S, Nicolis A, Rattazzi R (2006) Causality, analyticity and an IR obstruction to UV completion. JHEP 0610:014. doi:10.1088/1126-6708/2006/10/014

Babichev E, Mukhanov V, Vikman A (2008) k-Essence, superluminal propagation, causality and emergent geometry. JHEP 0802:101. doi:10.1088/1126-6708/2008/02/101

Berezhiani L, Chkareuli G, Gabadadze G (2013) Restricted Galileons

Burrage C, de Rham C, Heisenberg L, Tolley AJ (2012) Chronology protection in galileon models and massive gravity. JCAP 1207:004. doi:10.1088/1475-7516/2012/07/004 de Rham C, Gabadadze G (2010) Generalization of the Fierz-Pauli action. Phys Rev D 82:044020. doi:10.1103/PhysRevD.82.044020

de Rham C, Dvali G, Hofmann S, Khoury J, Pujolas O et al (2008) Cascading gravity: extending the Dvali-Gabadadze-Porrati model to higher dimension. Phys Rev Lett 100:251603. doi:10.1103/PhysRevLett.100.251603

de Rham C, Fasiello M, Tolley AJ (2013) Galileon duality

Deffayet C, Deser S, Esposito-Farese G (2009) Generalized Galileons: all scalar models whose curved background extensions maintain second-order field equations and stress-tensors. Phys Rev D 80:064015. doi:10.1103/PhysRevD.80.064015

Deffayet C, Deser S, Esposito-Farese G (2010) Arbitrary p-form Galileons. Phys Rev D 82:061501. doi:10.1103/PhysRevD.82.061501

Dvali G, Franca A, Gomez C (2012) Road signs for UV-completion

Dvali G, Giudice GF, Gomez C, Kehagias A (2011) UV-completion by classicalization. JHEP 1108:108. doi:10.1007/JHEP08(2011)108

Goon G, Hinterbichler K, Trodden M (2011) Stability and superluminality of spherical DBI galileon solutions. Phys Rev D 83:085015. doi:10.1103/PhysRevD.83.085015

Hinterbichler K, Trodden M, Wesley D (2010) Multi-field galileons and higher co-dimension branes. Phys Rev D 82:124018. doi:10.1103/PhysRevD.82.124018

Nicolis A, Rattazzi R, Trincherini E (2009) The Galileon as a local modification of gravity. Phys Rev D 79:064036. doi:10.1103/PhysRevD.79.064036

Padilla A, Saffin PM, Zhou S-Y (2010) Bi-galileon theory I: motivation and formulation. JHEP 1012:031. doi:10.1007/JHEP12(2010)031

Padilla A, Saffin PM, Zhou S-Y (2011) Bi-galileon theory II: phenomenology. JHEP 1101:099. doi:10.1007/JHEP01(2011)099

Vikman A (2013) Suppressing quantum fluctuations in classicalization. Europhys Lett 101:34001. doi:10.1209/0295-5075/101/34001

Part III
Quantum Corrections in Massive Gravity

Chapter 5
Quantum Corrections: Natural Versus Non-natural

The cosmological constant problem is an inveterate puzzle which resisted several decades of a considerable amount of effort devoted by numerous physicists. It reflects the enormous discrepancy between field theory predictions and observations. Similar puzzles are also encountered within the Standard Model of particle physics, for example the Higgs Hierarchy problem of why the Higgs mass is so small relative to the Planck scale. These hierarchies are puzzling as they do not seem to be protected without the help of new physics, such as supersymmetry. Solutions proposed to solve the Hierarchy problem are helpless to tackle the cosmological constant problem since they rely on physics at different scales. Technically natural tunings on the other hand are very common within the Standard model of particle physics. For example the electron mass m_e is much smaller than the electroweak scale, however is technically natural. In the limit $m_e \rightarrow 0$ there is an enhancement of the symmetry of the system, due to the recovery of chiral symmetry. The existence of this symmetry in the massless limit is enough to protect the electron mass from receiving large quantum corrections thanks to the 't Hooft naturalness argument ('t Hooft and Veltman 1974; 't Hooft 1980). Therefore quantum corrections will only give rise to a renormalization of the electron mass proportional to itself, and thus the hierarchy between the electron mass and the electroweak scale is technically natural.

In the case of the cosmological constant, λ, there is no symmetry recovered in the limit $\lambda \rightarrow 0$. The Einstein Hilbert action is invariant under general diffeomorphism and a cosmological constant can be included to this action without breaking this symmetry. Therefore the smallness of the cosmological constant is thus unnatural in the 't Hooft sense. Now an elegant way to tackle the cosmological constant problem would consist of introducing instead another technically natural small scale which could account for the late time acceleration. Thus, infrared modifications of General Relativity could give a hope to find a resolution to the cosmological constant problem. From a theoretical point of view, one might argue that a model with a technically natural small parameter, m, which describes the accelerated Universe is more elegant and offers theoretical advantages over the one in which dark energy is due to a technically-unnatural small cosmological constant. Of course this is a rather philosophical reasoning and can not be justified using a meaningful physical

L. Heisenberg, *Theoretical and Observational Consistency of Massive Gravity*,
Springer Theses, DOI 10.1007/978-3-319-18935-2_5

comparison as in the Bayesian evidence case. There is not such a thing as a measure of the beauty of a theoretical modeling. Anyways, we will follow the philosophy of preferring a theory based on technically natural parameters. Since massive gravity is one of the most relevant large scale modified theories of gravity, it gives a promising framework to describe the accelerated expansion of the late Universe without the need of fine-tuning. In order to give rise to a reliable accelerating expansion of the universe the graviton mass needs to be tuned to no much larger than the Hubble scale today. This tuning of the graviton mass m is of course in the same order than that of the vacuum energy $m^2 \approx 10^{-120} M_{Pl}^2$. Nevertheless, this tuning is expected to be technically natural since we recover a local symmetry in the limit $m \to 0$. However, there is a possible loophole in this reasoning: the $m \to 0$ limit is obviously discontinuous in the number of gravitational degrees of freedom and the presence of these extra polarizations for $m \neq 0$ deserves a special treatment in the context of naturalness. Nevertheless, this does not mean that the physical predictions of the theory are discontinuous. The presence of the Vainshtein mechanism in this model (Koyama et al. 2011a, b; Chkareuli and Pirtskhalava 2012; Sbisa et al. 2012), as well as general (Babichev et al. 2010) extensions of the Fierz-Pauli theory make most of the physical predictions identical to that of General Relativity in the massless limit. In this chapter we are not only interested in the question of whether or not the mass of the graviton is technically natural. It could be for instance that the quantum corrections yield contributions fulfilling the naturalness argument in the sense that they are proportional to the mass of the graviton but still detune the nice structure of the interaction potential which was chosen so as to guaranty the absence of ghost instabilities. If the detuning reintroduces the ghost at a scale much smaller than the Planck mass, than this turns the theory to be unreliable and unstable under quantum corrections. So we want to address the two essential questions in this chapter:

- Are the quantum corrections to the graviton mass technically natural, i.e. are they proportional to the mass of the graviton itself?
- Do the quantum corrections detune the specific structure of the potential interactions?

We will first study the quantum corrections in the decoupling limit ($m \to 0$, $M_{Pl} \to \infty$) of massive gravity and show that the graviton mass remains protected against quantum corrections. Since this analysis is very similar to the non-renormalization theorem in Galileon theories, we will first recapitulate the quantum corrections in Galileons. We will then move on to the full theory beyond the decoupling limit and first consider only the 1-loop contributions coming through the coupling to the matter fields and then extend the analysis to the 1-loop contributions with only virtual graviton running in the loops.

Furthermore, we shall emphasize that we are not addressing the old cosmological constant problem reflecting the robustness of the theoretical prediction of a large vacuum energy, which was already pointed out before the dark energy was even discovered. We assume that there exist some mechanism which provides a vanishing cosmological constant (it could be for instance due to degravitation which was the original idea and motivation of massive gravity). On top of this mechanism we study the naturalness of massive gravity.

5.1 Non-renormalization Theorem for the Galileon Theory

In the following we will demonstrate why the Galileon interactions are protected against quantum corrections. First of all, the shift and Galileon symmetry will prevent to generate local operators by loop corrections which breaks explicitly these symmetries. Nevertheless, quantum corrections might still generate local operators which are invariant under shift and Galileon transformations, either renormalizing Galileon interactions themselves or generating operators of higher derivative interactions. It is the non-renormalization theorem, which ensures that the Galileon interactions themselves are not renormalized at all and that the higher derivative operators are irrelevant corrections in the regime of validity of the effective field theory $\partial^n\pi \ll \Lambda^{n+1}$ for $n \geq 3$. We will show the non-renormalization theorem perturbatively at the Feynman diagram level by showing how each vertex in an arbitrary Feynman diagram gives rise to interactions with at least one more derivative. For this purpose, it is convenient to write the Galileon interactions 5.8 in the following form

$$\begin{aligned}
\mathcal{L}_2 &= \pi\mathcal{E}^{\mu\alpha\rho\sigma}\mathcal{E}^{\nu}{}_{\alpha\rho\sigma}\Pi_{\mu\nu}, \\
\mathcal{L}_3 &= \pi\mathcal{E}^{\mu\alpha\rho\sigma}\mathcal{E}^{\nu\beta}{}_{\rho\sigma}\Pi_{\mu\nu}\Pi_{\alpha\beta}, \\
\mathcal{L}_4 &= \pi\mathcal{E}^{\mu\alpha\rho\sigma}\mathcal{E}^{\nu\beta\gamma}{}_{\sigma}\Pi_{\mu\nu}\Pi_{\alpha\beta}\Pi_{\rho\gamma} \\
\mathcal{L}_5 &= \pi\mathcal{E}^{\mu\alpha\rho\sigma}\mathcal{E}^{\nu\beta\gamma\delta}\Pi_{\mu\nu}\Pi_{\alpha\beta}\Pi_{\rho\gamma}\Pi_{\sigma\delta}\,.
\end{aligned} \tag{5.1}$$

Without loss of generality, let us for a moment concentrate on the cubic Galileon interaction $\pi\mathcal{E}^{\mu\alpha\rho\sigma}\mathcal{E}^{\nu\beta}{}_{\rho\sigma}\Pi_{\mu\nu}\Pi_{\alpha\beta}$. We will consider an arbitrary Feynman diagram with this interaction at a given vertex (Fig. 5.1). We can contract an external helicity-0 particle π with momentum q_μ with the helicity-0 field coming without derivatives in this vertex from the cubic Galileon while letting the other two π-particles run in the loop with momenta k_μ and $(q+k)_\mu$. The contribution of this vertex to the scattering amplitude is

$$\mathcal{A} \propto \int \frac{d^4k}{(2\pi)^4} G_k G_{k+q} \mathcal{E}^{\mu\alpha\rho\sigma}\mathcal{E}^{\nu\beta}{}_{\rho\sigma}\, k_\mu k_\nu\, (q+k)_\alpha\, (q+k)_\beta \cdots , \tag{5.2}$$

where $G_k = k^{-2}$ is the Feynman massless propagator for the Galileon field. Now, it is a trivial observation that all the terms which are linear in the external momentum $\mathcal{E}^{\mu\alpha\rho\sigma}\mathcal{E}^{\nu\beta}{}_{\rho\sigma}k_\mu k_\nu k_\alpha q_\beta$ as well as all the contributions which are independent of it $\mathcal{E}^{\mu\alpha\rho\sigma}\mathcal{E}^{\nu\beta}{}_{\rho\sigma}k_\alpha k_\beta k_\mu k_\nu$ will cancel owing to the antisymmetric nature of the vertex (carried by the indices in the Levi-Civita symbols). Therefore, the only non-vanishing term will come in with at least two powers of the external Galileon field

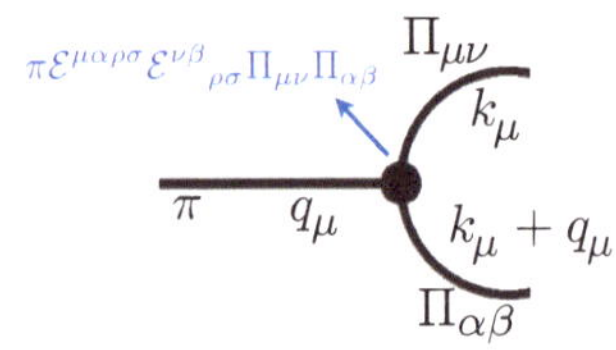

Fig. 5.1 An arbitrary Feynman diagram with the cubic Galileon vertex

with momentum $q_\alpha q_\beta$. This is the essence of the non-renormalization theorem of the Galileon interactions (5.1). In a similar way, if we contract the external leg with the derivative free field in vertices of quartic and quintic Galileon, the same argument straightforwardly leads to the same conclusion regarding the minimal number of derivatives on external fields. Therefore there is no counterterm which takes the Galileon form, and the Galileon interactions are hence not renormalized. This would mean that the Galileon coupling constants may be technically natural tuned to any value and remain radiatively stable. On the other hand, to understand why the generation of higher derivative interactions is harmless, bear in mind the fact that for fluctuations $\delta\pi$ on top of the background configuration, interactions do not arise at the scale Λ_3 but rather at the rescaled strong coupling scale $\tilde{\Lambda}_3 = \sqrt{Z}\Lambda_3$ (where Z is the modified kinetic matrix due to the background configuration as in 1.9, i.e. $Z \sim \left(1 + \frac{\partial^2\bar{\pi}_0}{\Lambda^3} + \frac{(\partial^2\bar{\pi}_0)^2}{\Lambda^6} + \cdots\right)$), which is much larger than Λ_3 within the strong coupling region. The higher interactions for fluctuations on top of the background configuration are hence much smaller than expected and their quantum corrections are therefore suppressed.

The non-renormalization theorem is not unique to the Galileon theory only. A similar non-renormalization theorem applies in the DBI scalar field models, in which a brane is embedded within a wrapped extra dimension

$$\begin{aligned}\mathcal{L}_{\text{DBI}} &= f(\pi)^4\left(1 - \sqrt{1 + f(\pi)^{-4}(\partial\pi)^2}\right) - V(\pi)\\ &= -\frac{1}{2}(\partial\pi)^2 + \frac{(\partial\pi)^4}{6f(\pi)^4} + \cdots - V(\pi)\end{aligned} \tag{5.3}$$

In a similar way as in the Galileon theory the field operator itself and its velocity can be considered to be large $\pi \sim f(\pi)$ and $\partial\pi \sim f(\pi)^2$ as long as the higher derivatives are suppressed $\partial^n\pi \ll f(\pi)^{n+1}$ with $n \geq 2$ and hence the quantum corrections are negligible.

In a straightforward manner one can apply the same philosophy to the de Sitter Galileons, which describe Galileon interactions on a de Sitter background which we introduced in Sect. 1.3. In this case, the de Sitter Galileon interactions are dressed with corrections coming from the Hubble parameter, schematically as $\partial\pi \rightarrow \partial\pi + \beta H\pi$, with a given constant β so that for instance, $(\partial\pi)^2\Box\pi \rightarrow (\partial\pi)^2\Box\pi + 6H^2\pi(\partial\pi)^2 - 8H^4\pi^3$. In this framework, the Hubble parameter plays a similar role as the gradient, $H \sim \partial$, so that the previous counting remains identical. As a consequence, the de Sitter Galileon interactions are again not renormalized, and the effective field theory remains under control as long as $\partial/\Lambda \sim H/\Lambda \ll 1$, i.e. $\partial^3\pi_{cl}/\Lambda^4 \sim H^3\pi_{cl}/\Lambda^4 \ll 1$ even around classical configurations with $\partial^2\pi_{cl}/\Lambda^3 \sim H^2\pi_{cl}/\Lambda^3 \sim 1$. In the next section we will explicitly show that the same non-renormalization theorem applies in the decoupling limit of massive gravity. Because of the antisymmetric structure of the interactions in the decoupling limit, the quantum corrections yield terms which belong to a different class of interactions and therefore do not renormalize the decoupling limit interactions. The proof works exactly in the same way as for the Galileons.

5.2 Non-renormalization Theorem in the Decoupling Limit of Massive Gravity

Before we demonstrate the non-renormalization theorem for the decoupling limit of massive gravity, lets recapitulate some of the important properties and equations of this limit. As we already explained in more detail in Chap. 2 the leading part of the ghost-free massive gravity is described by an action giving by the interactions of the helicity-2 and helicity-0 polarizations of the graviton in the limit $m \to 0$, $M_{Pl} \to \infty$ and $\Lambda_3 \equiv (M_{Pl}m^2)^{1/3} = \text{finite}$

$$\mathcal{L} = -\frac{1}{2}h^{\mu\nu}\mathcal{E}^{\alpha\beta}_{\mu\nu}h_{\alpha\beta} + h^{\mu\nu}\sum_{n=1}^{3}\frac{a_n}{\Lambda_3^{3(n-1)}}X^{(n)}_{\mu\nu}(\Pi)\,, \tag{5.4}$$

with the three matrices X's explicitly given by the following expressions in terms of $\Pi_{\mu\nu} = \partial_\mu\partial_\nu\pi$ and the Levi-Civita symbol $\mathcal{E}^{\mu\nu\alpha\beta}$

$$\begin{aligned}
X^{(1)}_{\mu\nu}(\Pi) &= \mathcal{E}_\mu{}^{\alpha\rho\sigma}\mathcal{E}_\nu{}^\beta{}_{\rho\sigma}\Pi_{\alpha\beta},\\
X^{(2)}_{\mu\nu}(\Pi) &= \mathcal{E}_\mu{}^{\alpha\rho\gamma}\mathcal{E}_\nu{}^{\beta\sigma}{}_\gamma\Pi_{\alpha\beta}\Pi_{\rho\sigma},\\
X^{(3)}_{\mu\nu}(\Pi) &= \mathcal{E}_\mu{}^{\alpha\rho\gamma}\mathcal{E}_\nu{}^{\beta\sigma\delta}\Pi_{\alpha\beta}\Pi_{\rho\sigma}\Pi_{\gamma\delta}\,.
\end{aligned} \tag{5.5}$$

The three matrices $X^{(1,2,3)}$ are respectively linear, quadratic and cubic in $\partial^2\pi$, so that the action involves operators up to quartic order in the fields. As you can see, the interactions are very similar to the Galileon interactions with the difference that the π field in front of the interactions in Eq. (5.1) is replaced by the $h_{\mu\nu}$ field. This decoupling limit comprises a massless spin-two field $h_{\mu\nu}$, and a scalar field π, which couple to each other via some dimension 4, 7, and 10 operators; the latter two are suppressed by powers of the dimensionful scale Λ_3. The interactions become strong at the energy scale $E \sim \Lambda_3$. In a similar way as in the Galileon theory, even though the interactions involve more then two derivatives on the scalar field π, the theory is ghost-free (de Rham and Gabadadze 2010; de Rham et al. 2011). It means that it propagates exactly 2 polarizations of the massless tensor field and exactly one massless scalar.

The theory contains the following symmetries:

- linearized diffeomorphisms, $h_{\mu\nu} \to h_{\mu\nu} + \partial_{(\mu}\xi_{\nu)}$, which represent a symmetry of the full non-linear decoupling limit action up to total derivative (i.e. including the interactions $h^{\mu\nu}X^{(n)}_{\mu\nu}$)
- (global) field-space Galilean transformations, $\pi \to \pi + b_\mu x^\mu + b$.

In difference to the Galileon theory the symmetry for the scalar field in this decoupling limit is an exact symmetry without total derivatives. Similarly as for Galileons, these symmetries protect the theory from a certain class of quantum corrections, which do not satisfy the symmetry. In the following, we will show explicitly how the operators of the decoupling limit of massive gravity remain protected against quantum corrections to all orders in perturbation theory. At the heart of this non-renormalization is the specific anti-symmetric structure of the interaction vertices containing two derivatives per scalar line, all contracted by the epsilon tensors. Then, it is not too difficult to show that the loop diagrams cannot induce any renormalization of the tree-level terms in (5.4). Conceptually, the non-renormalization appears because the tree-level interactions in the Lagrangian are diffeomorphism invariant up to total derivatives only. On the other hand, the variations of the Lagrangian with respect to the fields in this theory are exactly diffeomorphism invariant. Therefore, no Feynman diagram can generate operators that would not be diffeomorphism invariant, and the original operators that are diffeomorphism invariant only up to total derivatives stay unrenormalized. This becomes apparent by looking at the resumed Feynman diagrams, the one-particle irreducible effective action (1PI action)[1]

$$e^{-S_{\rm eff}(h_{ab},\pi)} = \int \mathcal{D}h\mathcal{D}\pi e^{-h\left(\frac{\delta^2 S}{\delta h^2}|_{\pi=\bar{\pi}}\right)h-\pi\left(\frac{\delta^2 S}{\delta \pi^2}|_{h_{ab}=\bar{h}_{ab}}\right)\pi} \qquad (5.6)$$

The expression on the left hand side fulfills the symmetry exactly because the right hand side contains the variation with respect to the fields which fulfill the symmetry exactly as the equations of motion. This is similar to non-renormalization of the Galileon operators (Luty et al. 2003; Nicolis and Rattazzi 2004; Nicolis et al. 2009), with diffeomorphism invariance replaced by galilean invariance which is also fulfilled up to total derivatives. Let us consider one-loop terms in the 1PI action. These are produced by an infinite number of one-loop diagrams with external h and/or π lines. The diagrams contain power-divergent terms, the log-divergent pieces, and finite terms. The power-divergent terms are arbitrary, and cannot be fixed without the knowledge of the UV completion. For instance, dimensional regularization would set these terms to zero. Alternatively, one could use any other regularization, but perform subsequent subtraction so that the net result in the 1PI action is zero. In contrast, the log divergent terms are uniquely determined and would have to be included in the 1PI action.

All the induced terms in the 1PI action would appear suppressed by the scale Λ_3, since the latter is the only scale in the effective field theory approach (including the

[1] Without loss of generality we assume for now that h and π are diagonalized.

scale of the UV cutoff). Moreover, due to the same specific structure of the interaction vertices that guarantees non-renormalization of (5.4), the induced terms will have to have more derivatives per field than the unrenormalized terms. Therefore at low energies, formally defined by the condition $(\partial/\Lambda_3) \ll 1$, the tree-level terms will dominate over the induced terms with the same number of fields, as well as over the induced terms with a greater number of fields and derivatives. This property clearly separates the unrenormalized terms from the induced ones, and shows that the theory (5.4) is a good effective field theory below the scale Λ_3.

As an alternative approach, one can rely on the decoupling limit to infer that the additional degrees of freedom do decouple in the massless limit, and the local symmetry is indeed recovered as $m \to 0$. Beyond this limit, the free parameters of the theory are expected to be renormalized, albeit by an amount that should vanish in the limit. As a result, quantum corrections to the three defining parameters of the full theory (namely the mass m and the two free coefficients $a_{2,3}$) are strongly suppressed. In particular, the graviton mass receives a correction proportional to itself (with a coefficient that goes as $\delta m^2/m^2 \sim \left(m/M_{Pl}\right)^{2/3}$), thus establishing the technical naturalness of the theory.

One should stress at this point that technical naturalness is not an exclusive property of ghost-free massive gravity. Even theories with the Boulware-Deser ghost, can be technically natural, satisfying the $\delta m^2 \propto m^2$ property (Arkani-Hamed et al. 2003). Besides the fact that the latter theories are unacceptable, there are two important distinctions between the theories with and without Boulware-Deser ghosts. These crucial distinctions can be formulated in the decoupling limit, which occurs at a much lower energy scale, $\Lambda_5 = (m^4 M_{Pl})^{1/4} \ll \Lambda_3$, if the theory propagates a Boulware-Deser ghost. In the latter case the classical part of the decoupling limit is not protected by a non-renormalization theorem. As a consequence: (a) quantum corrections in ghost-free theories are significantly suppressed with respect to those in the theories with the Boulware-Deser ghost, and (b) unlike a generic massive gravity, the non-renormalization guarantees that *any relative tuning of the parameters in the ghost-free theories, that is* m, a_2, a_3, *is technically natural.* The latter property makes any relation between the free coefficients of the theory stable under quantum corrections. For example, a particular ghost-free theory with the decoupling limit, characterized by the vanishing of all interactions in (5.4) has been studied due to its simplicity (e.g. see (Buchbinder et al. 2012) for one-loop divergences in that model). The non-renormalization of ghost-free massive gravity in this case guarantees that such a vanishing of the classical scalar-tensor interactions holds in the full quantum theory as well.

Using the antisymmetric structure of these interactions, we can follow roughly the same arguments as for Galileon theories to show the RG invariance of these parameters, Luty et al. (2003). The only possible difference may emerge due to the gauge invariance $h_{\mu\nu} \to h_{\mu\nu} + \partial_{(\mu}\xi_{\nu)}$, and consequently the necessity of gauge fixing for the tensor field. Working in e.g., the de Donder gauge, the relevant modification of the arguments is trivial: gauge invariance is Abelian, so the corresponding Faddeev-Popov ghosts are free and do not affect the argument in any way. Moreover, the gauge fixing term changes the graviton propagator, but as we shall see below, all the arguments that follow solely depend on the special structure of vertices and are hence independent of the exact structure of the propagator. With these arguments in mind, one can thus proceed with the proof of the non-renormalization of the theory without being affected by gauge invariance.

The scalar π only appears within interactions/mixings with the spin-2 field in (5.4). In order to associate a propagator with it, we have to diagonalize the quadratic lagrangian by eliminating the $h^{\mu\nu}X^{(1)}_{\mu\nu}(\Pi)$ term. Such a diagonalization gives rise to a kinetic term for π, as well as additional scalar self-interactions of the Galileon form (de Rham and Gabadadze 2010),

$$\mathcal{L} = -\frac{1}{2}h^{\mu\nu}\mathcal{E}^{\alpha\beta}_{\mu\nu}h_{\alpha\beta} + \frac{3}{2}\pi\Box\pi + \left(h^{\mu\nu} + \pi\,\eta^{\mu\nu}\right)\sum_{n=2}^{3}\frac{a_n}{\Lambda_3^{3(n-1)}}X(n)_{\mu\nu}(\Pi)\,, \quad (5.7)$$

(here the interactions of the form $\pi X^{(n)}(\Pi)$ are nothing else but the cubic and quartic Galileons).

In the special case when the parameter a_3 vanishes, all scalar-tensor interactions are redundant and equivalent to pure scalar Galileon self-interactions as we have seen in Chap. 2. This can be seen through the field redefinition (under which the S-matrix is invariant) $h_{\mu\nu} = \tilde{h}_{\mu\nu} + \pi\,\eta_{\mu\nu} - \frac{2a_2}{\Lambda_3^3}\partial_\mu\pi\partial_\nu\pi$. We then recover a decoupled spin-2 field, supplemented by the Galileon theory for the scalar of the form

$$\mathcal{L}_{\rm Gal} = -\frac{1}{2}\sum_{n=0}^{2}\frac{b_n}{\Lambda_3^{3n}}X^{(n)}_{\mu\nu}(\Pi)\,\partial^\mu\pi\partial^\nu\pi\,, \quad (5.8)$$

where the Galileon coefficients b_n are in one-to-one correspondence with a_n and $X^{(0)}_{\mu\nu} \equiv \eta_{\mu\nu}$. The non-renormalization of the theory (5.4) then directly follows from the analogous property of the Galileons. For $a_3 \neq 0$, such a redefinition is however impossible (de Rham and Gabadadze 2010). This can be understood by noting that the $h^{\mu\nu}X^{(3)}_{\mu\nu}$ coupling encodes information about the linearized Riemann tensor for $h_{\mu\nu}$, which can not be expressed through π on the basis of the lower-order equations of motion (Chkareuli and Pirtskhalava 2012).

We will now show that, similarly to what happens in the pure Galileon theories, any external particle comes along with at least two derivatives acting on it in the 1PI action, hence establishing the non-renormalization of the operators present in (5.4) (the only difference comes from the fact that instead of π in front of each interaction, we now have $h_{\mu\nu}$). Of course, we keep in mind that these operators are merely the leading piece of the full 1PI action, which features an infinite number of additional higher derivative terms. They however are responsible for most of the phenomenology that the theories at hand lead to, making the non-renormalization property essential (Fig. 5.2).

The absence of the ghost in these theories is tightly related to the antisymmetric nature of their interactions, which in turn guarantees their non-renormalization.

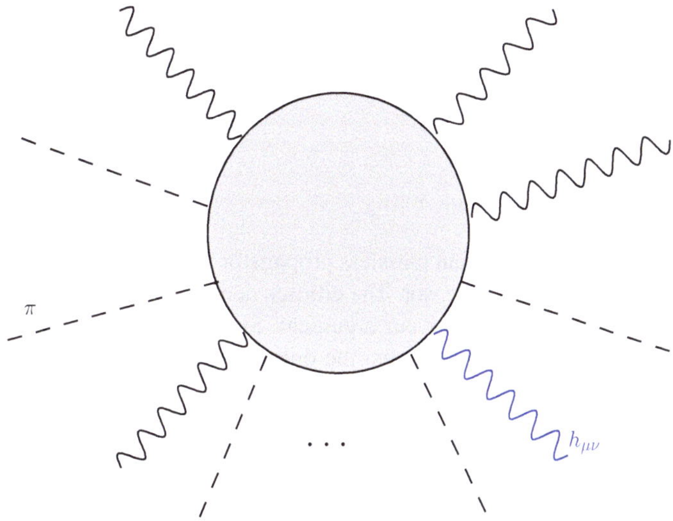

Fig. 5.2 An arbitrary 1PI diagram with gravitational degrees of freedom in the loop

The same reasoning applies to the construction of the Lovelock invariants. For example, in linearized GR, linearized diffemorphism tells us that the kinetic term can be written using the antisymmetric Levi–Cevita symbols as $\mathcal{E}^{\mu\nu\alpha\beta}\mathcal{E}^{\mu'\nu'}{}_{\alpha\beta}R_{\mu\nu\mu'\nu'}$, which ensures the non-renormalization. Notice that gauge invariance alone would still allow for a renormalization of the overall factor of the linearized Einstein–Hilbert term, which does not occur in the decoupling limit.

The key point is that any external particle attached to a diagram has at least two derivatives acting on it. This in turn implies that the operators generated are all of the form $(\partial^2\pi)^{n_1}(\partial^2 h)^{n_2}$, $n_1, n_2 \in \mathbb{N}$, and so are not of the same class as the original operators. This means that a_2 and a_3 are not renormalized. Furthermore, the new operators that appear in the 1PI are suppressed by higher powers of derivatives.

To be more precise, any external particle contracted with a field with two derivatives in a vertex contributes to a two-derivatives operator acting on this external particle—this is the trivial case. On the other hand, if we contract the external particles with fields without derivatives we could in principle generate operators with fewer derivatives. But now the antisymmetric structure of the interactions plays a crucial role. Without loss of generality consider for instance the interaction $V \supseteq h^{\mu\nu}\mathcal{E}_{\mu}{}^{\alpha\rho\gamma}\mathcal{E}_{\nu}{}^{\beta\sigma}{}_{\gamma}\Pi_{\alpha\beta}\Pi_{\rho\sigma}$, and contract an external helicity-2 particle with momentum p_μ with the helicity-2 field coming without derivatives in this vertex; the other two π-particles again run in the loop with momenta k_μ and $(p+k)_\mu$ (Fig. 5.3). The contribution of this vertex gives (de Rham et al. 2012)

$$\mathcal{A} \propto \int \frac{d^4k}{(2\pi)^4} G_k\, G_{k+p}\, f^{\mu\nu}\, \mathcal{E}_{\mu}{}^{\alpha\rho\gamma}\, \mathcal{E}_{\nu}{}^{\beta\sigma}{}_{\gamma}\, k_\alpha\, k_\beta\, (p+k)_\rho\, (p+k)_\sigma \cdots, \qquad (5.9)$$

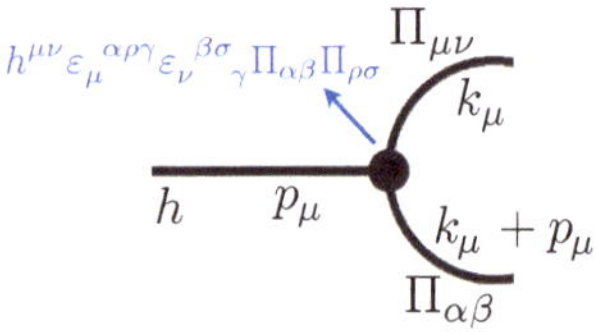

Fig. 5.3 An arbitrary Feynman diagram with the decoupling limit interaction $h^{\mu\nu}\mathcal{E}_\mu{}^{\alpha\rho\gamma}\mathcal{E}_\nu{}^{\beta\sigma}{}_\gamma\Pi_{\alpha\beta}\Pi_{\rho\sigma}$ at the vertex

where $G_k = k^{-2}$ is the Feynman massless propagator for the helicity-0 mode π, and $f^{\mu\nu}$ is the spin-2 polarization tensor. The ellipses denote the remaining terms of the diagram, which are irrelevant for our argument. Similarly to what happened in the case of the pure Galileon interactions, the only non-vanishing contribution to the scattering amplitude will come in with at least two powers of the external helicity-2 momentum $p_\rho p_\sigma$

$$\mathcal{A} \propto f^{\mu\nu}\mathcal{E}_\mu{}^{\alpha\rho\gamma}\mathcal{E}_\nu{}^{\beta\sigma}{}_\gamma\, p_\rho p_\sigma \int \frac{d^4k}{(2\pi)^4} G_k\, G_{k+p}\, k_\alpha\, k_\beta \cdots , \tag{5.10}$$

which in coordinate space corresponds to two derivatives on the external helicity-2 mode.

Thus any external leg coming out of the $h^{\mu\nu}X^{(2)}_{\mu\nu}$ vertex will necessarily have two or more derivatives on the corresponding field in the effective action. The same is trivially true for the $\pi X^{(2)}_{\mu\nu}$ vertex which corresponds to the pure Galileon interactions.

Similarly, if the external leg is contracted with the derivative-free field in vertices $h^{\mu\nu}X^{(3)}_{\mu\nu}$ and $\pi X^{(3)}_{\mu\nu}$, their contribution will always involve the external momentum p_μ and the loop momenta k_μ and k'_μ with the following structure,

$$\begin{aligned}\mathcal{A} \propto &\int \frac{d^4k d^4k'}{(2\pi)^8} G_k G'_k G_{k+k'+p} f^{\mu\nu}\mathcal{E}_\mu{}^{\alpha\rho\gamma}\mathcal{E}_\nu{}^{\beta\sigma\delta} k_\alpha k_\beta k'_\rho k'_\sigma \\ &\times (p+k+k')_\gamma (p+k+k')_\delta \cdots \\ \propto\; & f^{\mu\nu}\mathcal{E}_\mu{}^{\alpha\rho\gamma}\mathcal{E}_\nu{}^{\beta\sigma\delta} p_\gamma p_\delta \int \frac{d^4k d^4k'}{(2\pi)^8} G_k G_{k'} G_{k+k'+p}\, k_\alpha k_\beta k'_\rho k'_\sigma \cdots ,\end{aligned} \tag{5.11}$$

where the contraction on the Levi-Civita symbols is performed with either the graviton polarization tensor $f^{\mu\nu}$ or with $f^{\mu\nu} = \eta^{\mu\nu}$ depending on whether we are dealing with the vertex $h^{\mu\nu}X^{(3)}_{\mu\nu}$ or $\pi X^{(3)\mu}{}_\mu$. Similar arguments, as can be straightforwardly checked, lead to the same conclusion regarding the minimal number of derivatives on external fields for cases in which there are two external states coming out of these vertices (with the other two consequently running in the loops).

This completes the proof of the absence of quantum corrections to the scale Λ_3, the two parameters $a_{2,3}$, as well as to the spin-2 kinetic term and the scalar-tensor kinetic mixing in the theory defined by (5.4). This is the essence of the non-renormalization theorem in the decoupling limit of massive gravity: there are no quantum corrections to the two parameters a_2 and a_3, nor to the scale Λ_3. Moreover, the kinetic term of the helicity-2 mode is radiatively stable.

5.3 Implications for the Full Theory

In this section we will comment on the implications of the above emergent decoupling limit non-renormalization property for the full theory.

We have established in the previous section that in the decoupling limit the leading scalar-tensor part of the action does not receive quantum corrections in massive gravity: all operators generated by quantum corrections in the effective action have at least two extra derivatives compared to the leading terms, making the coefficients a_i invariant under the renormalization group flow. This in particular implies the absence of wave-function renormalization for the helicity-2 and helicity-0 fields in the decoupling limit. Moreover, the coupling with external matter fields goes as $\frac{1}{M_{Pl}} h_{\mu\nu} T^{\mu\nu}$ and thus is negligible as $M_{Pl} \to \infty$. The non-renormalization theorem is thus unaffected by external quantum matter fields in the decoupling limit.

The decoupling limit analysis of the effective action, much like the analogous nonlinear sigma models of non-Abelian spin-1 theories (Vainshtein and Khriplovich 1971), provides an important advantage over the full treatment (see Arkani-Hamed et al. (2003) for a discussion of these matters). In addition to being significantly simpler, the decoupling limit explicitly displays the relevant degrees of freedom and their (most important) interactions. In fact, as we will see below, we will be able to draw important conclusions regarding the magnitude of quantum corrections to the full theory based on the decoupling limit power counting analysis alone.

Now, whatever the renormalization of the specific coefficients α_i (and more generally, of any relative coefficient between terms of the form $[H^{\ell_1}] \cdots [H^{\ell_n}]$ in the graviton potential) is in the full theory, it has to vanish in the decoupling limit, since α_i are in one-to-one correspondence with the unrenormalized decoupling limit parameters a_i. Let us work in the unitary gauge, in which $H_{\mu\nu} = h_{\mu\nu}$, and for example look at quadratic terms in the graviton potential. We start with an action, the relevant part of which (in terms of the so-far dimensionless $h_{\mu\nu}$) is

$$\mathcal{L} \supset -\frac{1}{4} M_{Pl}^2 m^2 \left((1 + c_1)\, h_{\mu\nu}^2 - (1 + c_2)\, h^2 + \cdots \right), \tag{5.12}$$

where c_1 and c_2 are generated by quantum corrections after integrating out a small Euclidean shell of momenta and indices are assumed to be contracted with the flat metric. There is of course no guarantee that the two constants $c_{1,2}$ are equal, so they could lead to a detuning of the Fierz-Pauli structure and consequently to a ghost below the cutoff, unless sufficiently suppressed. Returning to the Stückelberg formalism, in terms of the canonically normalized fields

$$h_{\mu\nu} \to \frac{h_{\mu\nu}}{M_{Pl}}, \qquad \pi \to \frac{\pi}{M_{Pl} m^2} \tag{5.13}$$

the tree-level part (i.e. the one without c_1 and c_2) of the above Lagrangian would lead to the following scalar-tensor kinetic mixing in the decoupling limit (5.4)

$$\mathcal{L} \supset -h^{\mu\nu}\left(\partial_\mu \partial_\nu \pi - \eta_{\mu\nu} \Box \pi\right) + \cdots \,. \tag{5.14}$$

Now, from the decoupling limit analysis, we know that this mixing does not get renormalized. What does this imply for the renormalization of the graviton mass and the parameters of the potential in the full theory?

One immediate consequence of such non-renormalization is that in the decoupling limit, c_1 and c_2 both vanish. To infer the scaling of these parameters with M_{Pl}, let us look at the scalar-tensor interactions that arise beyond the decoupling limit. They are of the following schematic form[2]

$$\mathcal{L} \supset \sum_{n\geq 1,\,\ell\geq 0} \frac{f_{n,\ell}}{\Lambda_3^{3(\ell-1)}} h(\partial^2\pi)^\ell \left(\frac{h}{M_{Pl}}\right)^n \,, \tag{5.15}$$

i.e., they are all suppressed by an *integer* power of M_{Pl}^{-1} compared to vertices arising in the decoupling limit. Then, judging from the structure of these interactions, generically the non-renormalization theorem for the classical scalar-tensor action should no longer be expected to hold when considering the previous vertices present beyond the decoupling limit.

This implies that c_1 and c_2 generated by quantum corrections are of the form (remember that c_1 and c_2 are dimensionless)

$$c_{1,2} \sim \left(\frac{\Lambda_3}{M_{Pl}}\right)^k \,, \tag{5.16}$$

with k some positive integer $k \geq 1$, if the loops are to be cut off at the Λ_3 scale[3] (the fact that k needs to be an integer relies on the fact that the theory remains analytic beyond the decoupling limit). Taking the worst possible case (i.e., $k = 1$), one can directly read off the magnitude of the coefficients $c_{1,2}$,

$$c_{1,2} \lesssim \left(\frac{\Lambda_3}{M_{Pl}}\right) \sim \left(\frac{m}{M_{Pl}}\right)^{2/3} \,. \tag{5.17}$$

In terms of the quantum correction to the graviton mass itself, this implies (using simply $\delta m^2 = m^2\, \delta c_{1,2}$)

$$\delta m^2 \lesssim m^2 \left(\frac{m}{M_{Pl}}\right)^{2/3} \,, \tag{5.18}$$

[2] We are omitting here the part containing the helicity-1 interactions, which can uniquely be restored due to diff invariance of the helicity-2+helicity-1 system, and the U(1) invariance of the helicity-1+helicity-0 system.

[3] In this analysis, the graviton mass m is completely absorbed into Λ_3, and nothing special happens at the scale m as far as the strong coupling is concerned.

providing an explicit realization of technical naturalness for massive gravity.

Naively, this implies that the correction to the graviton mass goes like

$$\delta m^2 \sim \Lambda_3^2 \, (m/M_{Pl})^n, \tag{5.19}$$

where n is any positive number, since in this case the r.h.s. of the latter equation vanishes in the decoupling limit. This can be recast in the form $\delta m^2 \sim m^2 \, (m/M_{Pl})^{n-2/3}$ (up to the fact of course, that if the Foerz Pauli tuning is spoiled at $\mathcal{O}(1)$, m loses a physical interpretation of being the graviton mass), therefore for $0 < n \leq 2/3$ the theory would be badly non-technically natural. But n in fact should be larger than 2/3, as a consequence of the non-renormalization theorem. Indeed, taking the decoupling limit and computing quantum corrections should commute, so the non-renormalization of the leading decoupling limit action means the following: if one takes the original Lagrangian (5.12) and takes the decoupling limit (*by keeping exactly the same scalings for different helicities*) first by setting $c_1 = c_2 = 0$ and then by giving these constants their actual values computed from loops, one should get the same answer. This implies $c_1, c_2 \sim (m/M_{Pl})^k$, where $k > 0$. On the other hand, $n < 2/3$ would imply that c_1 and c_2 blow up in the Λ_3 decoupling limit, which is inconsistent with non-renormalization. This can be verified by looking at explicit operators, generated quantum mechanically. Although the couplings of the leading part of the decoupling limit action does not run in the effective theory, there certainly are operators of e.g. the form,

$$\frac{(\partial^2\pi)^n}{\Lambda_3^{3n-4}}, \tag{5.20}$$

in the Wilsonian action, the couplings of which do run with the renormalization group. The quadratic operator from the latter set is directly linked to the renormalization of the graviton mass.[4] Moreover, simply looking at the $n = 2$ operator from (5.20), one can directly read off the magnitude of the coefficients $c_{1,2}$,

$$c_{1,2} \sim \left(\frac{m}{M_{Pl}}\right)^{2/3}, \tag{5.21}$$

providing an explicit realization of technical naturalness for massive gravity.

One can extend these arguments to an arbitrary interaction in the effective potential. Consider a generic term of the following schematic form in the unitary gauge involving ℓ factors of the (dimensionless) metric perturbation

$$\mathcal{L} \supset M_{Pl}^2 m^2 \sqrt{-g} \, (\bar{c} + c) h^{\ell} \, . \tag{5.22}$$

[4] Of course, as noted above, one should be cautious about the meaning of "graviton mass", if $c_1 \neq c_2$ in (5.12), which one should anticipate to hold in the quantum theory. However, such a detuning of the Fierz-Pauli structure, as is well-known, does not spoil consistency of the effective theory if $c_{1,2} \ll 1$, which is true for the theory at hand.

Here indices are contracted with the full metric, $\bar{c}$ denotes the "classical" coefficient of the given term obtained from the ghost-free theory, and c is its quantum correction. Our task is to estimate the magnitude of c based on the non-renormalization of the decoupling limit scalar-tensor Lagrangian. Introducing back the Stückelberg fields through the replacement $h_{\mu\nu} \to H_{\mu\nu}$, and recalling the definition of different helicities (5.13), the quantum correction to the given interaction can be schematically written in terms of the various canonically normalized helicities as follows

$$\left(1+\frac{h}{M_{Pl}}+\cdots\right)^{1+\ell}\left(\frac{h}{M_{Pl}}+\frac{\partial A}{M_{Pl}m}+\frac{\partial^2\pi}{\Lambda_3^3}+\frac{\partial A\partial^2\pi}{M_{Pl}m\Lambda_3^3}+\frac{(\partial A)^2}{M_{Pl}^2m^2}+\frac{(\partial^2\pi)^2}{\Lambda_3^6}\right)^{\ell}.$$

The first parentheses denotes a schematic product of $\sqrt{-g}$ and ℓ factors of the inverse metric, needed to contract the indices. In the classical ghost-free massive gravity, the pure scalar self-interactions are carefully tuned to collect into total derivatives, projecting out the Boulware-Deser ghost. From the decoupling limit arguments, we know that quantum corrections do produce such operators, *e.g.* of the form $(\partial^2\pi)^{\ell}$, suppressed by the powers of Λ_3. This immediately bounds the magnitude of the coefficient c to be the same as for the $\ell = 2$ case

$$c \lesssim \left(\frac{\Lambda_3}{M_{Pl}}\right) \sim \left(\frac{m}{M_{Pl}}\right)^{2/3} . \tag{5.23}$$

Indeed, for c given by (5.23), we get $M_{Pl}^2 m^2 c \sim \Lambda_3^4$ and the upper bound on c is the same as that coming from the mass term renormalization. In the next chapter we will explicitly compute the quantum corrections beyond the decoupling limit and see whether or not these implications from the decoupling limit remain true in the full theory.

References

Arkani-Hamed N, Georgi H, Schwartz MD (2003) Effective field theory for massive gravitons and gravity in theory space. Annals Phys 305:96–118. doi:10.1016/S0003-4916(03)00068-X

Babichev E, Deffayet C, Ziour R (2010) The recovery of general relativity in massive gravity via the vainshtein mechanism. Phys Rev D 82:104008. doi:10.1103/PhysRevD.82.104008

Buchbinder IL, Pereira DD, Shapiro IL (2012) One-loop divergences in massive gravity theory. Phys Lett B 712:104–108. doi:10.1016/j.physletb.2012.04.045

Chkareuli G, Pirtskhalava D (2012) Vainshtein mechanism In Λ_3 - theories. Phys Lett B 713:99–103. doi:10.1016/j.physletb.2012.05.030

de Rham C, Gabadadze G (2010) Generalization of the fierz-pauli action. Phys Rev D 82:044020. doi:10.1103/PhysRevD.82.044020

de Rham C, Gabadadze G, Heisenberg L, Pirtskhalava D (2012) Non-renormalization and naturalness in a class of scalar-tensor theories. Phys Rev D 87

de Rham C, Gabadadze G, Tolley AJ (2011) Resummation of massive gravity. Phys Rev Lett 106:231101. doi:10.1103/PhysRevLett.106.231101 't Hooft G (1980) Naturalness, chiral symmetry, and spontaneous chiral symmetry breaking. NATO Adv Study Inst Ser B Phys 59:135

't Hooft G, Veltman MJG (1974) One loop divergencies in the theory of gravitation. Annales Poincare Phys Theor A20:69–94

Koyama K, Niz G, Tasinato G (2011a) Analytic solutions in non-linear massive gravity. Phys Rev Lett 107:131101. doi:10.1103/PhysRevLett.107.131101

Koyama K, Niz G, Tasinato G (2011b) Strong interactions and exact solutions in non-linear massive gravity. Phys Rev D 84:064033. doi:10.1103/PhysRevD.84.064033

Luty MA, Porrati M, Rattazzi R (2003) Strong interactions and stability in the DGP model. JHEP 0309:029

Nicolis A, Rattazzi R (2004) Classical and quantum consistency of the DGP model. JHEP 0406:059. doi:10.1088/1126-6708/2004/06/059

Nicolis A, Rattazzi R, Trincherini E (2009) The Galileon as a local modification of gravity. Phys Rev D 79:064036. doi:10.1103/PhysRevD.79.064036

Sbisa F, Niz G, Koyama K, Tasinato G (2012) Characterising vainshtein solutions in massive gravity. Phys Rev D 86:024033. doi:10.1103/PhysRevD.86.024033

Vainshtein AI, Khriplovich IB (1971) On the zero-mass limit and renormalizability in the theory of massive yang-mills field. Yad Fiz 13:198–211

Chapter 6
Renormalization Beyond the Decoupling Limit of Massive Gravity

In this chapter we study the naturalness of the graviton mass and whether or not the specific structure of the graviton gets detuned in the full theory beyond the decoupling limit, and compare our conclusions to the estimations coming from the decoupling limit analysis. More concrete, we will address the two essential questions.

- Do the overall parameters of the full theory get large quantum corrections? In order to have a viable cosmological phenomenon, the mass of the graviton needs to be tuned to a very small value. This tuning is as extreme as for the cosmological constant. Therefore the first essential question is whether this tuning is technically natural, meaning whether the graviton mass receives large quantum corrections. The results from the decoupling limit suggest a technically natural mass for the graviton and we would like to investigate this explicitly by computing the one-loop quantum corrections beyond the decoupling limit.
- How do the specific interactions that we need to avoid the ghost renormalize. For instance we need at linear order $c_1 h^2 + c_2 h^2_{\mu\nu}$ with $c1 = 1$ and $c_2 = -1$, but now if they renormalize differently, i.e. if the counter terms we need to add for h^2 and for $h^2_{\mu\nu}$ do not have the same tuning as the classical coefficients, then we would expect the ghost to come back to us at the quantum level. For example in the DGP model this combination is imposed upon us from 5 dimensional General Relativity and there we know that the 5 dimensional diffeomorphism invariance prevents us from having different quantum corrections, i.e. the 5 dimensional symmetry protects us against the reemergence of the ghost at the quantum level. Now in massive gravity, what we have to do, is either try to find a symmetry or compute some of the quantum corrections explicitly to see if the specific form of the interactions we need is maintained at the quantum level. Since the interactions of massive gravity are not protected by any known symmetry we do expect that the nice structure of the interactions will be detuned by quantum corrections. The question then becomes at which scale this detuning happens and whether the ghost reappears at a smaller scale.

For the quantum stability analysis, in this chapter we focus on quantum corrections arising at one-loop only, and assume for simplicity a massive scalar field as the

L. Heisenberg, *Theoretical and Observational Consistency of Massive Gravity*,
Springer Theses, DOI 10.1007/978-3-319-18935-2_6

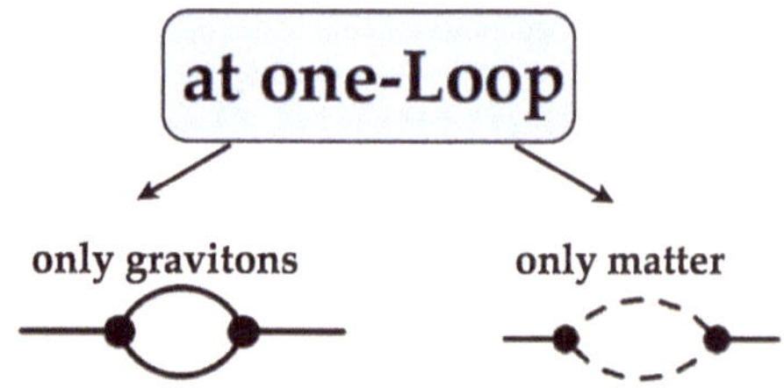

Fig. 6.1 At one loop there is either the matter fields (*dashed line*) running in the loop or the gravitons but not both simultaneously

external matter field. In particular this implies that either gravitons or matter field are running in the loops, but not *both* simultaneously (Fig. 6.1). Furthermore, by working in dimensional regularization, we discard any measure issues in the path integral related to field redefinitions which show up in power law divergences. We thus focus on logarithmic running results which are independent of this measure factor—in the language of field theory we concentrate on the runnings of the couplings.

Moreover, our aim is to study the stability of the graviton *potential* against quantum corrections, rather than the whole gravity action. As a result it is sufficient to address the diagrams for which the external graviton legs have zero momenta (i.e. , we focus on the IR limit of the runnings). This approach is complementary to the work developed by in Buchbinder et al. (2012) who used the Schwinger–DeWitt expansion of the one-loop effective action. This method allows one to obtain the Seeley–DeWitt coefficients associated with the curvature invariants generated by quantum corrections (see also Buchbinder et al. (2007)). Our approach differs in two ways. First, we introduce a covariant coupling to the matter sector and obtain the quantum corrections generated by matter loops. And second, we go beyond the minimal model investigated by Buchbinder et al. (2012) and study the quantum corrections to the full potential. As a by product, we do not focus on the radiatively generated curvature terms since these would also arise in GR and would therefore not be exclusive of theories of massive gravity. Finally, we use units for which $\hbar = 1$.

6.1 Quantum Corrections in the Metric Formulation

In this section we will perform the computation in the metric formulation. Even if the end results will not be different the computation will differ significantly whether we use the metric or the vielbein formulation. In any case, it offers a good consistency check to perform the computation in both languages and point out the differences. Starting with the metric formulation, we will show perturbatively that the one loop matter contributions give rise to a cosmological constant in the same way as in General Relativity. Unfortunately, the one loop graviton contributions detune the nice structure. In the next section we will perform the computations in the vielbein formulation and obtain consistent results.

Let us first compute the quantum corrections coming from the one matter loop in usual General Relativity. Our staring point is the action for gravity and a massive scalar field which plays the role of our source for gravity,

$$S = \int d^4x \left(\mathcal{L}_{GR} + \mathcal{L}_{matter}\right) \tag{6.1}$$

$$= \frac{M_{Pl}^2}{2} \int d^4x \sqrt{-g} R - \int d^4x \sqrt{-g} \left(\frac{1}{2} (\partial\chi)^2 + \frac{1}{2} M^2 \chi^2\right). \tag{6.2}$$

As we already pointed out, we are only interested in the effect of loops of matter on the graviton mass and thus expand this action to second order in the metric perturbations around flat space-time $g_{\mu\nu} = \eta_{\mu\nu} + \frac{2}{M_{Pl}} h_{\mu\nu}$ such that the full Lagrangian to second order in h is

$$\mathcal{L} = -h^{\mu\nu} \hat{\mathcal{E}}^{\alpha\beta}_{\mu\nu} h_{\alpha\beta} + \frac{1}{M_{Pl}} h_{\mu\nu} T^{\mu\nu} + \frac{1}{2M_{Pl}^2} h_{\mu\nu} h_{\alpha\beta} T^{\mu\nu\alpha\beta}, \tag{6.3}$$

where $\hat{\mathcal{E}}$ is the Lichnerowicz operator and $T_{\mu\nu}$ is the stress-energy tensor, whilst $T^{\mu\nu\alpha\beta}$ is its derivative with respect to the metric,

$$T_{\mu\nu} = \partial_\mu \chi \partial_\nu \chi - \frac{1}{2} \eta_{\mu\nu} \left((\partial\chi)^2 + M^2 \chi^2\right) \tag{6.4}$$

$$T_{\mu\nu\alpha\beta} = \partial_\mu \chi \partial_\nu \chi \eta_{\alpha\beta} + \partial_\alpha \chi \partial_\beta \chi \eta_{\mu\nu} - 4 \partial_{(\alpha} \chi \partial_{(\nu} \chi \eta_{\mu)\beta)} + \frac{1}{2} f_{\mu\nu\alpha\beta} \left((\partial\chi)^2 + M^2 \chi^2\right), \tag{6.5}$$

where we use the symmetrization convention $(a, b) = \frac{1}{2}(ab + ba)$ and in particular $4X_{(\mu(\alpha,\nu)\beta)} \equiv X_{\mu\alpha,\nu\beta} + X_{\mu\beta,\nu\alpha} + X_{\nu\alpha,\mu\beta} + X_{\nu\beta,\mu\alpha}$, and

$$f_{\mu\nu\alpha\beta} = \eta_{\mu(\alpha}\eta_{\nu\beta)} - \frac{1}{2} \eta_{\mu\nu} \eta_{\alpha\beta}. \tag{6.6}$$

With these conventions, the correctly normalized Feynman propagator for the graviton and the scalar field are

$$G_{\mu\nu\alpha\beta} = \langle h_{\mu\nu}(x_1) h_{\alpha\beta}(x_2)\rangle = f_{\mu\nu\alpha\beta} \int \frac{d^4k}{(2\pi)^4} \frac{e^{ik_\mu(x_1^\mu - x_2^\mu)}}{k^2 - i\varepsilon} \tag{6.7}$$

$$G_\chi = \langle \chi(x_1) \chi(x_2)\rangle = \int \frac{d^4k}{(2\pi)^4} \frac{e^{ik_\mu(x_1^\mu - x_2^\mu)}}{k^2 + M^2 - i\varepsilon}. \tag{6.8}$$

For simplicity, we will perform the computation in the Euclidean space. We go into Euclidean space by performing the Wick rotation $t \to -i\tau$, such that the Euclidean action is

$$S_E = -\int d^4x_E \left(\frac{(\partial\chi)^2}{2} + \frac{M^2}{2}\chi^2 + h^{\mu\nu}\left(\hat{\mathcal{E}}^{\alpha\beta}_{\mu\nu} h_{\alpha\beta} - \frac{T_{\mu\nu}}{M_{Pl}} - \frac{h^{\alpha\beta}T_{\mu\nu\alpha\beta}}{2M_{Pl}^2} \right)\right), \tag{6.9}$$

where all raising and lowering of the indices are now performed with the flat Euclidean space metric $\delta_{\mu\nu}$. Massive gravity on a Minkowski reference metric is thus mapped to massive gravity on a flat Euclidean reference metric δ_{ab} in Euclidean space.

6.1.1 Quantum Corrections in the Matter Loop

We want to focus on the renormalization of the graviton potential and not on its wave function or higher derivatives. For the potential renormalization it is sufficient to focus on the IR behaviour of these scattering amplitudes and set the momentum of any external leg to zero. Using these relations, we find that the 1-loop contribution to the tadpole (as represented on the first line of Fig. 6.2) is given by

$$\mathcal{A}^{(1pt)}_{\mu\nu} = \frac{1}{M_{Pl}} \int \frac{d^4k}{(2\pi)^4} \frac{k_\mu k_\nu - \frac{1}{2}\delta_{\mu\nu}(k^2+M^2)}{k^2+M^2} = \frac{1}{4}\frac{M^4}{M_{Pl}} J_{M,1}\delta_{\mu\nu}. \tag{6.10}$$

where

$$J_{M,1} = \frac{1}{M^4}\int \frac{d^4k}{(2\pi)^4}\frac{k^2}{k^2+M^2}, \tag{6.11}$$

as quoted in Appendix A.3.

Now turning to the 2-point correlation function, the 1-loop contribution arising for the cubic vertex $h_{\mu\nu}T^{\mu\nu}$ is given by

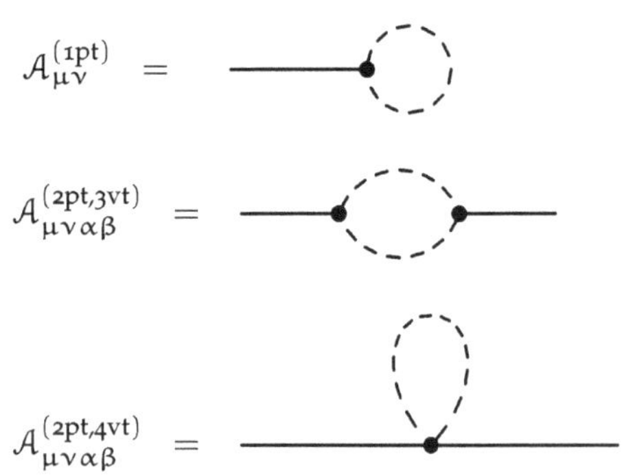

Fig. 6.2 1-loop matter contributions to the tadpole and 2-point correlation function

$$\begin{aligned}
\mathcal{A}^{(2pt,3vt)}_{\mu\nu\alpha\beta} &= \frac{(-1)^2\,(2\cdot 2)}{2!M_{Pl}^2}\int\frac{d^4k}{(2\pi)^4}\frac{k_\mu k_\nu k_\alpha k_\beta - \frac{1}{2}\left(\delta_{\mu\nu}k_\alpha k_\beta + \delta_{\alpha\beta}k_\mu k_\nu\right)(k^2+M^2)}{\left(k^2+M^2\right)^2}\\
&= \frac{2M^4}{M_{Pl}^2}\left[\frac{1}{24}J_{M,2}\left(2\delta_{\mu\alpha}\delta_{\nu\beta}+\delta_{\mu\nu}\delta_{\alpha\beta}\right) - \frac{1}{4}J_{M,1}\delta_{\mu\nu}\delta_{\alpha\beta}\right]\\
&= \frac{1}{4}\frac{M^4}{M_{Pl}^2}J_{M,1}\left(2\delta_{\mu\alpha}\delta_{\nu\beta}-\delta_{\mu\nu}\delta_{\alpha\beta}\right) = \frac{1}{2}\frac{M^4}{M_{Pl}^2}J_{M,1}\,f_{\mu\nu\alpha\beta},
\end{aligned} \tag{6.12}$$

where the terms in bracket in the first line are combinatory factors and already on the first line we ignored the terms that would cancel in dimensional regularization.

Finally focusing on the contribution from the quartic vertex, $h_{\mu\nu}h_{\alpha\beta}T^{\mu\nu\alpha\beta}$, we get

$$\begin{aligned}
\mathcal{A}^{(2pt,4vt)}_{\mu\nu\alpha\beta} &= \frac{(-1)\,(2)}{2M_{Pl}^2}\int\frac{d^4k}{(2\pi)^4}\frac{4\delta_{(\mu(\alpha}k_{\beta)}k_{\nu)} - \left(\delta_{\mu\nu}k_\alpha k_\beta + \delta_{\alpha\beta}k_\mu k_\nu\right)}{k^2+M^2}\\
&= -\frac{M^4}{M_{Pl}^2}J_{M,1}f_{\mu\nu\alpha\beta}.
\end{aligned} \tag{6.13}$$

The total 1-loop contribution to the 1-point and 2-point functions are thus given by

$$\mathcal{A}^{(1pt)}_{\mu\nu} = \frac{1}{4}\frac{M^4}{M_{Pl}}J_{M,1}\delta_{\mu\nu} \tag{6.14}$$

$$\mathcal{A}^{(2pt)}_{\mu\nu\alpha\beta} = \mathcal{A}^{(2pt,3vt)}_{\mu\nu\alpha\beta} + \mathcal{A}^{(2pt,4vt)}_{\mu\nu\alpha\beta} = -\frac{1}{2}\frac{M^4}{M_{Pl}^2}J_{M,1}\,f_{\mu\nu\alpha\beta}. \tag{6.15}$$

This corresponds to the following counter-terms at the level of the action

$$\mathcal{L}_{CT} = -\left(\mathcal{A}^{(1pt)}_{\mu\nu}h^{\mu\nu} + \frac{1}{2}\mathcal{A}^{(2pt)}_{\mu\nu\alpha\beta}h^{\mu\nu}h^{\alpha\beta}\right) \tag{6.16}$$

$$= -\frac{M^4}{4}J_{M,1}\left(\frac{1}{M_{Pl}}[h] + \frac{1}{2M_{Pl}^2}\left([h]^2 - 2[h^2]\right)\right) \tag{6.17}$$

$$= -\frac{M^4}{4}J_{M,1}\left(\sqrt{-g}-1\right) + \mathcal{O}\left(h^3/M_{Pl}^3\right), \tag{6.18}$$

so we see that the effective potential generated at 1-loop corresponds precisely to a cosmological constant with scale given by the scalar field mass M. This is precisely what causes the cosmological constant problem: any massive particle of mass M contributes to the vacuum energy proportional to M^4.

It is actually easier to derive the previous results by performing first a change of variable: $\psi = \sqrt{-g}\chi$. Since we are only interested in potential corrections to the graviton action and not in derivative corrections, any terms involving derivatives of g are irrelevant for this study (vanish in the IR limit), and the matter Lagrangian reduces to (keeping the same conventions as before)

$$\mathcal{L}_{\text{matter}} = -\int d^4x\sqrt{-g}\left(\frac{1}{2}(\partial\chi)^2 + \frac{1}{2}M^2\chi^2\right)$$
$$= -\int d^4x\left(\frac{g^{\mu\nu}\partial_\mu\psi\partial_\nu\psi}{2} + \frac{M^2}{2}\psi^2\right) + ((\partial g)\text{ corrections}).$$

The only place where the graviton interacts with matter is then through the kinetic term $g^{\mu\nu}\partial_\mu\psi\partial_\nu\psi$, and we then simply have the following interactions between gravity and the matter field ψ

$$\mathcal{L}_{h\psi} = \left(h^{\mu\nu} - 2(h^2)^{\mu\nu} + 4(h^3)^{\mu\nu}\right)\partial_\mu\psi\partial_\nu\psi, \tag{6.19}$$

with $(h^2)^{\mu\nu} = h^{\mu\alpha}h_\alpha{}^\nu$ and similarly for $(h^3)^{\mu\nu}$. Using these vertices we can then compute the n-point graviton scattering amplitudes. For simplicity we use the notation $\mathcal{A}^{(n\,pt)} = \mathcal{A}^{(n\,pt)}_{a_1b_1\cdots a_nb_n}h^{a_1b_1}\cdots h^{a_nb_n}$ such that the contributions are

$$\mathcal{A}^{1pt} = \frac{1}{M_{Pl}}\int\frac{d^4k}{(2\pi)^4}\frac{h^{\mu\nu}k_\mu k_\nu}{(k^2+M^2)} = \frac{1}{4}\frac{M^4}{M_{Pl}}J_{M,1}[h] \tag{6.20}$$

$$\mathcal{A}^{(2pt,3vt)} = \frac{2}{M_{Pl}^2}\int\frac{d^4k}{(2\pi)^4}\frac{h^{\mu\nu}h^{\alpha\beta}k_\mu k_\nu k_\alpha k_\beta}{(k^2+M^2)^2}$$
$$= \frac{1}{12}\frac{M^4}{M_{Pl}^2}J_{M,2}\left(2[h^2] + [h]^2\right) \tag{6.21}$$

$$\mathcal{A}^{(2pt,4vt)} = -\frac{4}{M_{Pl}^2}\int\frac{d^4k}{(2\pi)^4}\frac{(h^2)^{\mu\nu}k_\mu k_\nu}{(k^2+M^2)} = -\frac{M^4}{M_{Pl}^2}J_{M,1}[h^2] \tag{6.22}$$

so once again

$$\mathcal{A}^{(2pt)} = \mathcal{A}^{(2pt,3vt)} + \mathcal{A}^{(2pt,4vt)} = \frac{1}{4}\frac{M^4}{M_{Pl}^2}J_{M,1}\left([h]^2 - 2[h^2]\right) \tag{6.23}$$

and we can keep playing the same game for higher n-point functions, for instance the three-point scattering amplitude is given by (Fig. 6.3)

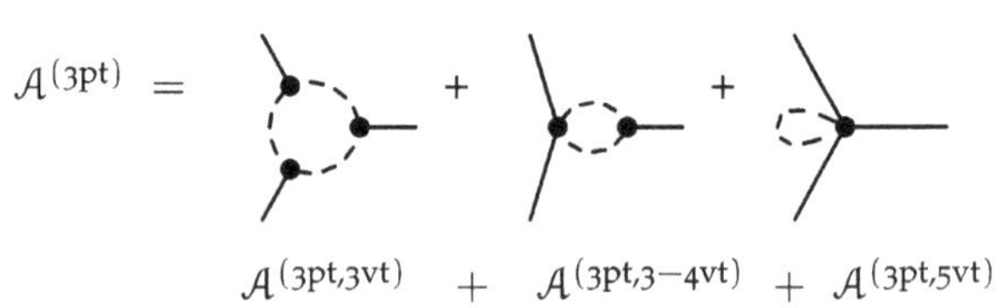

Fig. 6.3 1-loop matter contributions to the 3-point function

$$
\begin{aligned}
\mathcal{A}^{(3\mathrm{pt},3\mathrm{vt})} &= \frac{M^4}{3!M_{Pl}^3}(3!4.2)\frac{1}{192}\left([h]^3+6[h][h^2]+8[h^3]\right)J_{M,3} \\
\mathcal{A}^{(3\mathrm{pt},3-4\mathrm{vt})} &= -\frac{2.2M^4}{2!M_{Pl}^3}(3!.2)\frac{1}{24}\left([h^2][h]+2[h^3]\right)J_{M,2} \\
\mathcal{A}^{(3\mathrm{pt},5\mathrm{vt})} &= \frac{4M^4}{M_{Pl}^3}(3!)\frac{1}{4}[h^3]J_{M,1}
\end{aligned}
\tag{6.24}
$$

so that the total 3-point function goes as

$$
\mathcal{A}^{(3\mathrm{pt})} = \frac{1}{4}\frac{M^4}{M_{Pl}^3}J_{M,1}\left([h]^3-6[h][h^2]+8[h^3]\right), \tag{6.25}
$$

which is precisely the correct combination that describes a cosmological constant,

$$
\begin{aligned}
\mathcal{L}_{CT} &= -\left(\mathcal{A}^{(1\mathrm{pt})}+\frac{1}{2}\mathcal{A}^{(2\mathrm{pt})}+\frac{1}{3!}\mathcal{A}^{(3\mathrm{pt})}\right) \\
&= -\frac{M^4}{4}J_{M,1}\left(\frac{1}{M_{Pl}}[h]+\frac{1}{2M_{Pl}^2}\left([h]^2-2[h^2]\right)\right. \\
&\quad \left. +\frac{1}{6M_{Pl}^3}\left([h]^3-6[h][h^2]+8[h^3]\right)\right) \\
&= -\frac{M^4}{4}J_{M,1}\left(\sqrt{-g}-1\right)+\mathcal{O}\left(h^4/M_{Pl}^4\right),
\end{aligned}
\tag{6.26}
$$

The computation for the one-loop matter contributions was so far for the usual General Relativity, which gave rise to counter terms in form of a cosmological constant, which is at the origin of the cosmological constant problem. We now move onto the 1-loop effect of matter when the graviton is massive. The only modification comes from the graviton propagator, which is now given by

$$
G^{\mathrm{massive}}_{\mu\nu\alpha\beta} = \langle h_{\mu\nu}(x_1)h_{\alpha\beta}(x_2)\rangle = \tilde{f}_{\mu\nu\alpha\beta}\int\frac{d^4k}{(2\pi)^4}\frac{e^{ik_\mu(x_1^\mu-x_2^\mu)}}{k^2+m^2}, \tag{6.27}
$$

with

$$
\tilde{f}_{\mu\nu\alpha\beta} = \left(\tilde{\eta}_{\mu(\alpha}\tilde{\eta}_{\nu\beta)}-\frac{1}{3}\tilde{\eta}_{\mu\nu}\tilde{\eta}_{\alpha\beta}\right) \quad \text{and} \quad \tilde{\eta}_{\mu\nu} = \eta_{\mu\nu}+\frac{k_\mu k_\nu}{m^2}, \tag{6.28}
$$

i.e. its polarization is now no longer proportional to $\eta_{\mu(\alpha}\eta_{\beta)\nu}-\frac{1}{2}\eta_{\mu\nu}\eta_{\alpha\beta}$. However as we have seen in the massless case, to that level of the computation, considering only loops of matter, the graviton propagator has no effect on the computation of the effective potential and these effects are thus irrelevant. Not surprisingly, the one loop contributions in Massive Gravity are identical to General Relativity. The reason for that is that the graviton mass only affects the graviton propagator and not the vertices (as we are keeping the same coupling to matter as we do in General Relativity, i.e. a covariant coupling). Since loops only involve the matter field propagator and are independent to the graviton mass, there is no contributions from the graviton mass to these quantum corrections, and we recover exactly the same result as in General Relativity. So matter loops (at the 1-loop level) only give rise to a cosmological constant as in (6.26) but do not affect the structure of the graviton potential.

6.1.2 Quantum Corrections in the Graviton Loop

We now focus on quantum corrections arising from graviton loop (focusing first on the corrections from the potential to the potential). We will see that if we only consider effects from the potential, the quantum corrections to the mass are going as $\delta m^2 \sim m^4/M_{Pl}^2 \sim 10^{-120}\, m^2$ and are hence highly suppressed (the naturalness argument is fully present). However a key aspect to consider is whether or not the special structure of the potential (which is essential for the absence of ghost) is preserved under quantum corrections. We will organize the quantum corrections under powers of α_3 and α_4.

6.1.2.1 Quantum Corrections from Potential $\mathcal{U}_4$

We will now compute the one-loop graviton contributions coming from the potential interactions. Lets start with the potential $\mathcal{U}_4$, which was given by

$$\mathcal{L}_4 = \alpha_4 m^2 M_{Pl}^2 \sqrt{g}\, \mathcal{E}^{abcd} \mathcal{E}_{\alpha\beta\gamma\delta}\, \mathcal{K}^\alpha_a\, \mathcal{K}^\beta_b\, \mathcal{K}^\gamma_c\, \mathcal{K}^\delta_d \tag{6.29}$$

so that expanding it to leading order in $h_{\mu\nu}$ whilst keeping the notation $g_{\mu\nu} = \eta_{\mu\nu} + \frac{2}{M_{Pl}} h_{\mu\nu}$, we have

$$\mathcal{L}_4 = \alpha_4 \frac{m^2}{M_{Pl}^2}\, \mathcal{E}^{abcd} \mathcal{E}_{\alpha\beta\gamma\delta}\, h^\alpha_a\, h^\beta_b\, h^\gamma_c\, h^\delta_d + \mathcal{O}(h^5). \tag{6.30}$$

Notice that since the potential $\mathcal{U}_4$ starts at quartic order in h, it does not renormalize the tadpole at one-loop (or at linear order in α_4) and so there can be no contribution to the cosmological constant at this level. Nevertheless, this potential leads to quantum corrections to the 2-point function which go as

$$\begin{aligned}
\mathcal{A}^{(2pt,4vt)}_{\alpha_4} &= \mathcal{A}^{(2,4vt)}_{\mu\nu\alpha\beta} h^{\mu\nu} h^{\alpha\beta} \\
&= (4.3)\alpha_4 \frac{m^2}{M_{Pl}^2} h_{a\alpha} h_{b\beta} \int \frac{d^4k}{(2\pi)^4} \frac{\mathcal{E}^{abcd}\mathcal{E}_{\alpha\beta\gamma\delta}\, \tilde{f}_{c\gamma d\delta}}{\left(k^2+M^2\right)} \qquad (6.31) \\
&= -12\alpha_4 \frac{1}{M_{Pl}^2} h_{a\alpha} h_{b\beta} \int \frac{d^4k}{(2\pi)^4} \frac{1}{\left(k^2+M^2\right)} \\
&\quad \times \frac{5}{3}\left[k^a k^\beta \delta^{\alpha b} + k^b k^\alpha \delta^{a\beta} - k^a k^\alpha \delta^{b\beta} - k^b k^\beta \delta^{a\alpha}\right] \\
&= -10\alpha_4 \frac{m^4}{M_{Pl}^2} J_{m,1} \left([h^2] - [h]^2\right), \qquad (6.32)
\end{aligned}$$

using dimensional regularization. This is precisely of the Fierz-Pauli structure which means that this one-loop correction to the mass from terms linear in α_4 do preserve the structure of the potential. This are good news: the counter terms to the two point function coming from the quartic potential $\mathcal{U}_4$ are in form of the Fierz-Pauli structure (Fig. 6.4).

Now still at one loop and linear order in α_4, we expect a correction to the three point function from the terms going as $\alpha_4 \frac{m^2}{M_{Pl}^3} h^5$ etc. to all orders, and we should check that they combine to give terms which are simply a superposition of the three potentials $\mathcal{U}_{2,3,4}$ which would correspond to a renormalization of $\alpha_{2,3,4}$ but not a change of the structure.

Lets have an explicit look to the contribution to the three point function coming from h^5 terms. The interactions at linear order in α_4 and fifth order in h are expressed as follows (Fig. 6.5)

$$\begin{aligned}
\mathcal{L}^{5^{th}}_4 &= \frac{\alpha_4}{16} \frac{m^2}{M_{Pl}^3} \Big(-6 h^c_a h^{ab} h^d_b h^e_c h_{de} + 5([h^2]+[h]^2) h^e_c h^{cd} h_{de} \\
&\quad - 5[h]^3[h^2] + [h]^5\Big) \qquad (6.33)
\end{aligned}$$

$\mathcal{A}^{(2pt,4vt)}_{\alpha_4}$ =
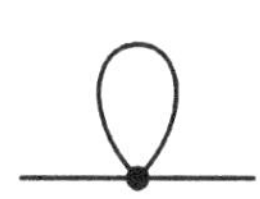

Fig. 6.4 One-loop contribution to the 2-point correlation function from the potential $\mathcal{U}_4$

$\mathcal{A}^{(3pt),5vt}_{\alpha_4}$ =

Fig. 6.5 1-loop contribution to the 3-point function linear in α_4

The contribution of these interactions to the three point function is of the Fierz-Pauli structure which means that this one-loop correction to the mass do **still** preserve the structure of the potential. Thus, the counter terms to the three point function coming from the quartic potential $\mathcal{U}_4$ are also in form of the Fierz-Pauli structure

$$\mathcal{A}_{\alpha_4}^{(3pt,5vt)} = -\alpha_4 \frac{25}{32} \frac{m^4}{M_{Pl}^3} J_{m,1}[h] \left([h^2] - [h]^2\right), \tag{6.34}$$

We have seen that the one-loop graviton quantum corrections coming from the quartic potential $\mathcal{U}_4$ give rise to contributions which do preserve the nice structure of the potential. These are promising results and motivate to study the quantum corrections coming from the cubic potential $\mathcal{U}_3$.

6.1.2.2 Quantum Corrections from Potential $\mathcal{U}_3$

We now look at the corrections to the tadpole and 2pt function arising at linear order in α_3 (at 1-loop). In difference to the quartic potential, the cubic potential can contribute to the tadpole and therefore could in principle yield a contribution in form of a cosmological constant. We expand again the cubic potential

$$\mathcal{L}_3 = \alpha_3 m^2 M_{Pl}^2 \sqrt{g}\, \mathcal{E}^{abcd} \mathcal{E}_{\alpha\beta\gamma d}\, \mathcal{K}_a^\alpha \mathcal{K}_b^\beta \mathcal{K}_c^\gamma \tag{6.35}$$

in terms of the metric perturbations

$$\mathcal{L}_3 = \frac{\alpha_3 m^2}{M_{Pl}} \mathcal{E}^{abcd} \mathcal{E}^{\alpha\beta\gamma}{}_d \left((1 + \frac{h}{M_{Pl}}) h_{a\alpha} h_{b\beta} h_{c\gamma} - \frac{9}{2M_{Pl}} h_{a\rho} h^\rho_\alpha h_{b\beta} h_{c\gamma} \right) + \mathcal{O}(h^5). \tag{6.36}$$

The cubic interactions in h from the cubic potential lead to a contribution to the tadpole at linear order in α_3 in the following form

$$\mathcal{A}_{\alpha_3}^{(1pt)} = (3)\alpha_3 \frac{m^2}{M_{Pl}} h_{a\alpha} \int \frac{d^4k}{(2\pi)^4} \frac{\tilde{f}_{b\beta c\gamma}}{\left(k^2 + m^2\right)} \mathcal{E}^{abcd} \mathcal{E}^{\alpha\beta\gamma}{}_d \tag{6.37}$$

$$= -\frac{15}{2} \frac{\alpha_3}{M_{Pl}} [h] \int \frac{d^4k}{(2\pi)^4} \frac{k^2 + 2m^2}{\left(k^2 + m^2\right)} \tag{6.38}$$

$$= \frac{15}{2} \alpha_3 \frac{m^4}{M_{Pl}} J_{m,1}[h]. \tag{6.39}$$

Let us now focus on the contributions to the 2-point function which are linear in α_3. They arise from the quartic interactions in (6.36) and there are 5 possible contractions (2 from the $hh_{a\alpha}h_{b\beta}h_{c\gamma}$ terms and 3 from the term going as $h_{a\rho}h^{\rho}_{\alpha}h_{b\beta}h_{c\gamma}$). Combining all these possibilities, we get

$$\begin{aligned}\mathcal{A}^{(2pt)}_{\alpha_3} &= \alpha_3 \frac{m^2}{M_{Pl}^2}\mathcal{E}^{abcd}\mathcal{E}^{\alpha\beta\gamma}{}_{d}\int \frac{d^4k}{(2\pi)^4}\frac{1}{(k^2+m^2)}\Big[(2\cdot 3)hh_{a\alpha}\tilde{f}_{b\beta c\gamma} \\ &\quad -\frac{9}{2}\Big((2)h^{\sigma}_{a}h_{\sigma\alpha}\tilde{f}_{b\beta c\gamma}+(8)h^{\sigma}_{a}h_{b\beta}\tilde{f}_{\sigma\alpha c\gamma}+(2)h_{b\beta}h_{c\gamma}\tilde{f}_{a\sigma}{}^{\sigma}{}_{\alpha}\Big) \\ &\quad +(2\cdot 3)h_{a\alpha}h_{b\beta}\tilde{f}_{c\gamma\sigma}{}^{\sigma}\Big]. \end{aligned} \tag{6.40}$$

Performing these contractions and using the identities in (A.20), we get

$$\begin{aligned}\mathcal{A}^{(2pt)}_{\alpha_3} &= \frac{-\alpha_3}{m^2M_{Pl}^2}\int\frac{d^4k}{(2\pi)^4}\frac{1}{(k^2+m^2)}\Big[\Big(26m^4+13k^2m^2+2k^4\Big)[h]^2 \\ &\quad -2\Big(28m^4+14k^2m^2+k^4\Big)[h^2]\Big] \\ &= \frac{15}{2}\alpha_3\frac{m^4}{M_{Pl}^2}J_{m,1}\Big([h]^2-2[h^2]\Big), \end{aligned} \tag{6.41}$$

which combined with (6.74), gives rise to precisely to the correct combination for the cosmological constant. Thus, the quantum corrections to the tadpole and 2pt function arising at linear order in α_3 combine into a contribution for the cosmological constant. These are excellent news. Not only the contributions coming from the quartic potential preserve the nice structure of the potential but also the contributions coming from the cubic potential do preserve it as well. These promising results motivate the exploration of quantum corrections to next leading order in the parameters. Lets have a quick look at the one-loop contributions to the 2-point function coming from the cubic potential at the quadratic order in α_3. There are two types of diagrams contributing to the 2pt functions at quadratic order in α_3 which we denote as $\mathcal{A}^{(2pt,1)}_{\alpha_3}$ and $\mathcal{A}^{(2pt,2)}_{\alpha_3}$ as in Fig. 6.6.

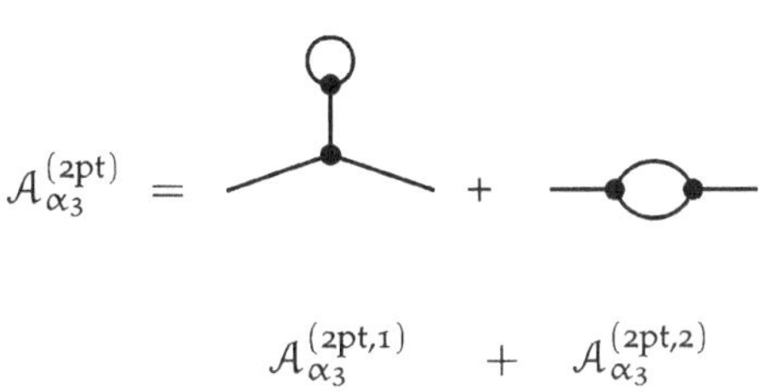

Fig. 6.6 1-loop contributions to the 2-point function at quadratic order in α_3

First of all, the contribution from the first diagram $\mathcal{A}_{\alpha_3}^{(2pt,1)}$ is

$$\begin{aligned}\mathcal{A}_{\alpha_3}^{(2pt,1)} &= \frac{\alpha_3^2}{2}\frac{m^4}{M_{Pl}^2} h_{aa'} h_{bb'} \mathcal{E}^{abcd} \mathcal{E}^{a'b'c'}{}_{d} \mathcal{E}^{\alpha\beta\gamma\delta} \mathcal{E}^{\alpha'\beta'\gamma'}{}_{\delta} \tilde{f}^{q=0}_{cc'\gamma\gamma'} \\ &\times \int \frac{d^4k}{(2\pi)^4} \frac{\tilde{f}_{\alpha\alpha'\beta'\beta'}}{(k^2+m^2)} \\ &\propto \frac{m^4}{M_{Pl}^2} \alpha_3^2 J_{m,1} \left([h^2] - [h]^2\right), \end{aligned} \tag{6.42}$$

and thus has precisely the correct structure of the Fierz-Pauli mass term. Inspired by the result of the first contribution, now when turning to the second contribution, we get

$$\begin{aligned}\mathcal{A}_{\alpha_3}^{(2pt,2)} &= \frac{\alpha_3^2}{2}(3\cdot 3\cdot 2\cdot 2)\frac{m^4}{M_{Pl}^2} h_{aa'} h_{\alpha\alpha'} \mathcal{E}^{abcd} \mathcal{E}^{a'b'c'}{}_{d} \mathcal{E}^{\alpha\beta\gamma\delta} \mathcal{E}^{\alpha'\beta'\gamma'}{}_{\delta} \\ &\times \int \frac{d^4k}{(2\pi)^4} \frac{f_{bb'\beta\beta'} f_{cc'\gamma\gamma'}}{(k^2+m^2)^2} \\ &= \frac{5}{2}\frac{m^4}{M_{Pl}^2} \alpha_3^2 J_{m,1} \left(7[h]^2 + 8[h^2]\right), \end{aligned} \tag{6.43}$$

which unfortunately does not correspond to anything nice and we failed to combine it into some contributions to the cosmological constant and Fierz-Pauli interaction. Even if all the quantum corrections do indeed have a natural scaling in the sense that they are proportional to the mass of the graviton and suppressed by the Planck mass and so proving the technical naturalness anticipated from the decoupling limit analysis, the nice potential structure to avoid ghost instabilities seem to be broken by quantum corrections. We also checked the quantum corrections coming from the quartic potential at the quadratic order in α_4 and we were finding similar disappointing results. The detuning of the potential at the quantum level seems to be an unavoidable feature. The question we should be then pursuing is at which scale the lowest detuning operator appears which would correspond to the scale at which the ghost would reappear. Before doing that, we will first recalculate the quantum corrections in the vielbein language and continue the analysis of the scale of the detuning in this language, which will facilitate the computation.

6.2 Quantum Correction in the Vielbein Language

In this section we would like to perform the computations in the vielbein language since in most of the cases they are easier to perform. First of all, we will review the tree level ghost-free covariant non-linear theory of massive gravity in the vielbein in the for us most convenient notation. We will first concentrate on the one-loop contributions arising from the coupling to external matter fields. Similarly as in the previous section using the metric language, we will be able to show that they only imply a running of the cosmological constant and of no other potential terms for the graviton. However, there will be a crucial difference: the perturbative expansion of the determinant of the metric in the vielbein language is finite, such that the Feynman diagrams only yield contributions up to the quartic order. We will also perform the one-loop effective action from the matter fields and confirm the same result in form of a cosmological constant. We will than move on and discuss the one-loop contributions from the gravitons themselves, and illustrate that whilst these destabilize the special structure of the potential, this detuning is irrelevant below the Planck scale. We then push the analysis further and show that even if the background configuration is large, as should be the case for the Vainshtein mechanism to work (1972), this will redress the one-loop effective action in such a way that the detuning remains irrelevant below the Planck scale.

6.2.1 *Ghost-Free Massive Gravity in the Vielbein Inspired Variables*

In this subsection we will recapitulate the ghost-free interactions in the theory of massive gravity in the vielbein inspired variables. Our starting point will be the results coming from the decoupling limit and to investigate the way quantum corrections affect the general structure of the potential in the vielbein language and compare these with the metric formulation results.

The presence of a square-root in the ghost-free realization of massive gravity makes its expression much more natural in the vielbein language (Groot et al. 2007; Chamseddine and Mukhanov 2011; Hinterbichler and Rosen 2012; Deffayet et al. 2013) (see also Gabadadze et al. 2013; Ondo and Tolley 2013). In the vielbein formalism, the ghost-free potential is polynomial and at most quartic in the vielbein fields. To make use of this natural formulation, we will work throughout this chapter in a 'symmetric-vielbein inspired language' where the metric is given by

$$g_{ab} = \left(\bar{\gamma}_{ab} + \frac{v_{ab}}{M_{Pl}}\right)^2 \equiv \left(\bar{\gamma}_{ac} + \frac{v_{ac}}{M_{Pl}}\right)\left(\bar{\gamma}_{db} + \frac{v_{db}}{M_{Pl}}\right)\delta^{cd}, \qquad (6.44)$$

where $\bar{g}_{ab} = \bar{\gamma}^2_{ab} = \bar{\gamma}_{ac}\bar{\gamma}_{bd}\delta^{cd}$ is the background metric, and v_{ab} plays the role of the fluctuations. We stress that the background metric $\bar{g}_{ab}$ need not be flat, even though the reference metric f_{ab} will be taken to be flat throughout this study, $f_{ab} = \delta_{ab}$.[1]

[1] In the Euclidean version of massive gravity both the dynamical metric $g_{\mu\nu}$ and the reference metric $f_{\mu\nu}$ have to be 'Euclideanized', $g_{\mu\nu} \rightarrow g_{ab}$ and $f_{\mu\nu} = \eta_{\mu\nu} \rightarrow \delta_{ab}$.

In this language, when working around a flat background metric, $\bar{\gamma}_{ab} = \delta_{ab}$, the normal fluctuations about flat space are expressed in terms of v as

$$g_{ab} - \delta_{ab} = \frac{2}{M_{Pl}} v_{ab} + \frac{1}{M_{Pl}^2} v_{ac} v_{bd} \delta^{cd}. \tag{6.45}$$

The conversion from g_{ab} to v_{ab} is a field redefinition that will contribute a measure term in the path integral. This generates power law divergent corrections to the action which, since we will work in dimensional regularization, can be ignored. This reflects the fact that the physics is independent of such field redefinitions, and only the logarithmic runnings are physically meaningful for the purposes of our study.

In this section our goal is to go beyond the non-renormalization argument in the decoupling limit reviewed above and investigate the quantum corrections in the full non-linear theory written in terms of the vielbein variables. We choose to work in the unitary gauge in which the Stückelberg fields vanish and $\Phi^a = X^\alpha \delta^a{}_\alpha$. The fluctuations $v_{\mu\nu}$ encode all the five physical degrees of freedom if it is massive (the two helicity-± 2, the two helicity-± 1 and the helicity-0 modes), and only the two helicity-± 2 modes if it is massless. To remind ourselves and to prevent to scroll back and forth, we summarize the important formulas. The Feynman propagator for the massless graviton is given by

$$G^{(\text{massless})}_{abcd} = \langle v_{ab}(x_1) v_{cd}(x_2) \rangle = f^{(0)}_{abcd} \int \frac{d^4k}{(2\pi)^4} \frac{e^{ik\cdot(x_1 - x_2)}}{k^2}, \tag{6.46}$$

in which $x_{1,2}$ are the Euclidean space coordinates, and where the polarization structure is given by

$$f^{(0)}_{abcd} = \delta_{a(c}\delta_{bd)} - \frac{1}{2}\delta_{ab}\delta_{cd}. \tag{6.47}$$

Here $\delta_{a(c}\delta_{bd)} \equiv \frac{1}{2}\delta_{ac}\delta_{bd} + \frac{1}{2}\delta_{ad}\delta_{bc}$. For the massive graviton, on the other hand, the corresponding Feynman propagator is given by

$$G^{(\text{massive})}_{abcd} = \langle v_{ab}(x_1) v_{cd}(x_2) \rangle = f^{(m)}_{abcd} \int \frac{d^4k}{(2\pi)^4} \frac{e^{ik\cdot(x_1 - x_2)}}{k^2 + m^2}, \tag{6.48}$$

with the polarization structure

$$f^{(m)}_{abcd} = \left(\tilde{\delta}_{a(c}\tilde{\delta}_{bd)} - \frac{1}{3}\tilde{\delta}_{ab}\tilde{\delta}_{cd} \right) \quad \text{where} \quad \tilde{\delta}_{ab} = \delta_{ab} + \frac{k_a k_b}{m^2}. \tag{6.49}$$

Notice that the polarization of the massive graviton is no longer proportional to $f^{(0)}_{abcd}$. Consequently when we take the massless limit, $m \to 0$, we do not recover the GR limit, which is at the origin of the vDVZ-discontinuity (van Dam and Veltman 1970; Zakharov 1970),

$$\lim_{m \to 0} f^{(m)}_{abcd} \neq f^{(0)}_{abcd} ,$$

which we had commented about in Sect. 1.2. At first sight this might be worrisome since on solar system and galactic scales gravity is in very good agreement with GR. Nevertheless, on these small scales, the effects of massive gravity can be cloaked by the Vainshtein mechanism (1972), where the crucial idea is to decouple the additional modes from the gravitational dynamics via nonlinear interactions of the helicity-0 graviton. The success of the Vainshtein mechanism relies on derivative interactions, which cause the helicity-0 mode to decouple from matter on short distances, whilst having observational signatures on larger scales. The implementation of the Vainshtein mechanism was so far at the classical level. In this section we will explicitly study how the Vainshtein mechanism acts at the quantum level, and how the quantum corrections do not diverge in the limit when $m \to 0$, even though the propagator (6.27) does.

We focus on the IR behaviour of the loop corrections. Starting with loops of matter, we will see that the peculiar structure in (6.112) has no effect on the computation of the quantum corrected effective potential at one-loop in the vielbein language. In a similar way as in the metric formulation in the previous section, this is because at one-loop the matter field and the graviton cannot both be simultaneously propagating in the loops if we consider only the contributions to the graviton potential. As a result, the quantum corrections are equivalent to those in GR. Only once we start considering loops containing virtual gravitons will the different polarization and the appearance of the mass in the propagator have an impact on the results. Furthermore, the graviton potential induces new vertices which also ought to be considered.

In terms of the 'vielbein-inspired' perturbations, the ghost-free potential becomes polynomial in v_{ab},

$$\mathcal{L} = \frac{M_{Pl}^2}{2} \sqrt{g} R - \frac{1}{4} M_{Pl}^2 m^2 \sum_{n=2}^{4} \frac{1}{M_{Pl}^n} \tilde{\alpha}_n \tilde{\mathcal{U}}_n[v]. \tag{6.50}$$

In unitary gauge the potential above is fully defined by

$$\tilde{\mathcal{U}}_2[h] = \mathcal{E}^{abcd} \mathcal{E}^{a'b'}{}_{cd} v_{aa'} v_{bb'} \tag{6.51}$$

$$\tilde{\mathcal{U}}_3[h] = \mathcal{E}^{abcd} \mathcal{E}^{a'b'c'}{}_{d} v_{aa'} v_{bb'} v_{cc'} \tag{6.52}$$

$$\tilde{\mathcal{U}}_4[h] = \mathcal{E}^{abcd} \mathcal{E}^{a'b'c'd'} v_{aa'} v_{bb'} v_{cc'} v_{dd'} , \tag{6.53}$$

where $\mathcal{E}^{abcd}$ represents the fully antisymmetric Levi-Cevita *symbol* (and not tensor, so in this language $\mathcal{E}^{abc}{}_{d} = \delta_{dd'} \mathcal{E}^{abcd'}$, for example, carries no information about the metric).

The first coefficient is fixed, $\tilde{\alpha}_2 = 1$, whereas the two others are free. They relate to the two free coefficients of de Rham et al. (2011b), as $\tilde{\alpha}_3 = -2(1+\alpha_3)$ and $\tilde{\alpha}_4 = -2(\alpha_3+\alpha_4) - 1$ (where α_3 and α_4 are respectively the coefficients of the potential $\mathcal{U}_3$ and $\mathcal{U}_4$ in that language).[2] The absence of ghost-like pathologies is tied to the fact that, when expressed in terms of π uniquely, (6.51)–(6.53) are total derivatives.

6.2.2 Quantum Corrections from Matter Loops in the Vielbein Language

In the previous section, we had seen how the non-renormalization theorem prevents large quantum corrections from arising in the decoupling limit of massive gravity. Since there the coupling to external matter fields was suppressed by the Planck scale these decouple completely when we take $M_{Pl} \to \infty$ limit.

Here, we will again keep the Planck scale, M_{Pl}, finite and look at the contributions from matter loops and investigate their effect on the structure of the graviton potential in the vielbein formulation. Again for definiteness, we consider gravity coupled to a scalar field χ of mass M and study one-loop effects. When focusing on the one-loop 1PI for the graviton potential, there can be no mixing between the graviton and the scalar field inside the loop. Furthermore, since we are interested in the corrections to the graviton potential, we only assume graviton zero momentum for the external legs. We still use dimensional regularization so as to focus on the running of the couplings, which are encoded by the logarithmic terms.

Our starting point is the Lagrangian for massive gravity (6.50) to which we add a real scalar field χ of mass M,

$$S = \int d^4x \left(\mathcal{L}_{mGR} + \mathcal{L}_{matter}\right), \tag{6.54}$$

with

$$\mathcal{L}_{matter} = -\sqrt{g}\left(\frac{1}{2}g^{ab}\partial_a\chi\partial_b\chi + \frac{1}{2}M^2\chi^2\right). \tag{6.55}$$

Note the sign difference due to the fact that this is the Euclidean action. The Feynman propagator for the scalar field reads

$$G_\chi = \langle\chi(x_1)\chi(x_2)\rangle = \int \frac{d^4k}{(2\pi)^4}\frac{e^{ik.(x_1-x_2)}}{k^2+M^2}. \tag{6.56}$$

---

[2] As in de Rham et al. (2011b), this 2-parameter family of potential is the one for which there is no cosmological constant nor tadpole.

The mixing between the scalar field and the graviton is encoded in (6.55) and is highly non-linear. Before proceeding any further we again perform the following change of variables for the scalar field

$$\chi \to (g)^{-1/4}\psi, \tag{6.57}$$

where $g \equiv \det\{g_{ab}\}$, so that the matter Lagrangian is now expressed as

$$\mathcal{L}_{\text{matter}} = \frac{g^{cd}}{2}\left(\partial_c\psi - \frac{1}{4}\psi g^{ab}\partial_c g_{ab}\right)\left(\partial_d\psi - \frac{1}{4}\psi g^{pq}\partial_d g_{pq}\right) + \frac{M^2\psi^2}{2}. \tag{6.58}$$

Since we will only be considering zero momenta for the external graviton legs, we may neglect the terms of the form ∂g. As a result, the relevant action for computing the matter loops in given by

$$S_{\text{matter}} = \int d^4x\left(\frac{1}{2}g^{cd}\partial_c\psi\partial_d\psi + \frac{1}{2}M^2\psi^2\right). \tag{6.59}$$

In what follows we will compute the one-loop effective action (restricting ourselves to a scalar field in the loops only) and show explicitly that the interactions between the graviton and the scalar field lead to the running of the cosmological constant, but not of the graviton potential. This comes as no surprise since inside the loops the virtual scalar field has no knowledge of the graviton mass and thus behaves in precisely the same way as in GR, leading to a covariant one-loop effective action. When it comes to the potential, the only operator it can give rise to which is covariant is the cosmological constant. We show this result explicitly in the one-loop effective action, and then present it in a perturbative way, which will be more appropriate when dealing with the graviton loops. This also provides a nice consistency check with the result obtained in the metric formulation.

6.2.2.1 One-Loop Effective Action

The one-loop effective action $S_{1,\text{eff}}(g_{ab}, \psi)$ is given by

$$e^{-S_{1,\text{eff}}(\bar{g}_{ab},\bar{\psi})} = \int \mathcal{D}\Psi e^{-\Psi^i(S_{ij}(\bar{g}_{ab},\bar{\psi}))\Psi^j}, \tag{6.60}$$

where Ψ_i is a placeholder for all the fields, $\Psi_i = \{g_{ab}, \psi\}$, and S_{ij} is the second derivative of the action with respect to those fields,

$$S_{ij}(\bar{g}_{ab}, \bar{\psi}) \equiv \left.\frac{\delta^2 S}{\delta\Psi^i\delta\Psi^j}\right|_{g_{ab}=\bar{g}_{ab},\psi=\bar{\psi}}. \tag{6.61}$$

Here $\bar{g}_{ab}$ and $\bar{\psi}$ correspond to the background quantities around which the action for fluctuations is expanded. Since we are interested in the graviton potential part of the one-loop effective action, we may simply integrate over the scalar field and obtain

$$e^{-S_{1,\text{eff}}^{(\text{matter-loops})}(\bar{g}_{ab})} = \int \mathcal{D}\psi e^{-\psi\left(\frac{\delta^2 S}{\delta\psi^2}|_{g_{ab}=\bar{g}_{ab},\psi=\bar{\psi}}\right)\psi}. \tag{6.62}$$

We therefore recover the well-known Coleman–Weinberg effective action,

$$S_{1,\text{eff}}^{(\text{matter-loops})}(\bar{g}_{ab}) = \frac{1}{2}\log\det\left(\frac{\delta^2 S}{\delta\psi^2}\bigg|_{g_{ab}=\bar{g}_{ab}}\right) = \frac{1}{2}\text{Tr}\log\left(\frac{\delta^2 S}{\delta\psi^2}\bigg|_{g_{ab}=\bar{g}_{ab}}\right). \tag{6.63}$$

Going into Fourier space this leads to

$$\begin{aligned}
\mathcal{L}_{1,\text{eff}}^{(\text{matter-loops})}(\bar{g}_{ab}) &= \frac{1}{2}\int \frac{d^4k}{(2\pi)^4}\log\left(\frac{1}{2}\bar{g}^{ab}k_a k_b + \frac{1}{2}M^2\right) \\
&= \frac{1}{2}\sqrt{\bar{g}}\int \frac{d^4\tilde{k}}{(2\pi)^4}\log\left(\frac{1}{2}\delta^{ab}\tilde{k}_a\tilde{k}_b + \frac{1}{2}M^2\right) \\
&\supset \frac{M^4}{64\pi^2}\sqrt{\bar{g}}\log(\mu^2),
\end{aligned} \tag{6.64}$$

where μ is the regularization scale and we restrict our result to the running piece. From the first to the second equality, we have performed the change of momentum $k_a \to \tilde{k}_a$ such that $g^{ab}k_a k_b = \delta^{ab}\tilde{k}_a\tilde{k}_b$. From this analysis, we see directly that the effect of external matter at one-loop is harmless on the graviton potential. This is no different from GR, since the scalar field running in the loops is unaware of the graviton mass, and the result is covariant by construction. This conclusion is easily understandable in the one-loop effective action (however, when it comes to graviton loops it will be harder to compute the one-loop effective action non-perturbatively and we will perform a perturbative analysis instead). For consistency, we apply a perturbative treatment to the matter fields as well in the next subsection and comment on the differences to the perturbative treatment used in the metric formulation in the previous section.

Higher Loops—Before moving on to the perturbative argument, we briefly comment on the extension of this result to higher loops. Focusing on matter loops only then additional self-interactions in the matter sector ought to be included. Let us consider, for instance, a $\lambda\chi^3$ coupling. The matter Lagrangian will then include a new operator of the form $\mathcal{L} \supseteq \lambda\sqrt{g}\chi^3 = \lambda g^{-1/4}\psi^3$, where $g = \det\{g_{ab}\}$. At n-loops, we have n integrals over momentum, and $2(n-1)$ vertices $\lambda g^{-1/4}\psi^3$, so the n-loop effective action reads symbolically

$$S_n^{(\text{matter-loops})}(\bar{g}_{ab}) = \frac{\lambda^{2(n-1)}}{g^{(n-1)/2}}\int \frac{d^4k_1\cdots d^4k_n}{(2\pi)^{4n}}\mathcal{F}_n\left(k_1^2,\ldots,k_n^2,M^2\right), \tag{6.65}$$

where $\mathcal{F}_n$ is a scalar function of the different momenta $k_j^2 = \bar{g}^{ab} k_{ja} k_{jb}$.[3] As a result one can perform the same change of variables as used previously, $k_j \to \tilde{k}_j$, with $k_j^2 = \delta^{ab} \tilde{k}_{ja} \tilde{k}_{jb} \equiv \tilde{k}_j^2$. This brings n powers of the measure $\sqrt{\bar{g}}$ down so that the n-loop effective action is again precisely proportional to $\sqrt{\bar{g}}$

$$S_n^{(\mathrm{matter-loops})}(\bar{g}_{ab}) = \frac{\lambda^{2(n-1)}}{g^{(n-1)/2}} g^{n/2} \left[\int \frac{d^4\tilde{k}_1 \cdots d^4\tilde{k}_n}{(2\pi)^{4n}} \mathcal{F}_n \left(\tilde{k}_1^2, \ldots, \tilde{k}_n^2, M^2 \right) \right]$$
$$\propto \sqrt{g}\ \overset{2}{\lambda}(n-1) M^{6-2n} \log \mu . \qquad (6.66)$$

The integral in square brackets is now completely independent of the metric $\bar{g}_{ab}$ and the n-loop effective action behaves as a cosmological constant. The same result holds for any other matter self-interactions. Once again this result is not surprising as this corresponds to the only covariant potential term it can be.

6.2.2.2 Perturbative Approach

In the previous subsection we have shown how at one-loop external matter fields only affect the cosmological constant and no other terms in the graviton potential. For consistency we show how this can be seen perturbatively in the vielbein-inspired perturbations about flat space, as defined in Eq. (6.45). Including all the interactions between the graviton and the matter field, but ignoring the graviton self-interactions for now, the relevant action is then

$$S = \int d^4x \left\{ v^{ab} \left[\hat{\mathcal{E}}^{cd}_{ab} + \frac{1}{2} M^2 \left(\delta^c_a \delta^d_b - \delta_{ab} \delta^{cd} \right) \right] v_{cd} \right.$$
$$\left. + \frac{1}{2} \sum_{n \geq 0} (-1)^n (n+1) (\hat{v}^{ab})^n \partial_a \psi \partial_b \psi + \frac{1}{2} M^2 \psi^2 \right\}, \qquad (6.67)$$

where $v_{ab} \equiv M_{Pl} \hat{v}_{ab}$ is the canonically normalized helicity-2 mode. Raising and lowering of the indices is now performed with respect to the flat Euclidean space metric, δ_{ab}, since we are working perturbatively. Note that we are using the notation $(\hat{v}^{ab})^2 \equiv \hat{v}^{ac} \hat{v}_c^{\ b}$.

We now calculate the one-loop matter contribution to the n-point graviton scattering amplitudes. For simplicity of notation, we again define the scattering amplitudes as

$$\mathcal{A}^{(n\,pt)} \equiv \mathcal{A}^{(n\,pt)}_{a_1 b_1 \cdots a_n b_n} \hat{v}^{a_1 b_1} \ldots \hat{v}^{a_n b_n} .$$

[3] Even if different momenta k_j contract one can always reexpress them as functions of k_j^2, following a similar procedure to what is presented in Appendix A.3.

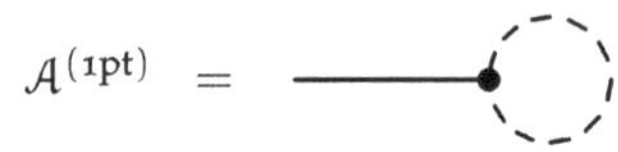

Fig. 6.7 Contribution to the graviton tadpole from a matter loop. *Dashes* denote the matter field propagator, whereas *solid lines* denote the graviton. This convention will be adopted throughout the paper

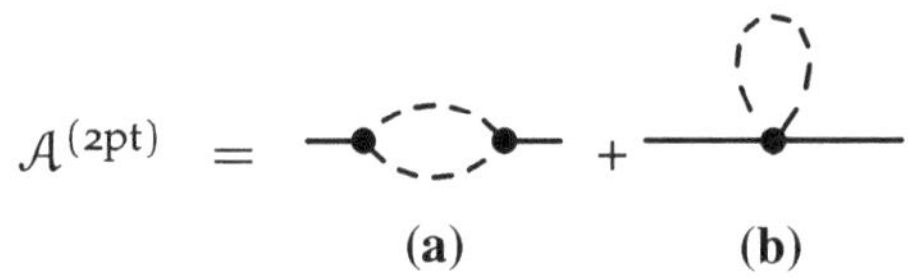

Fig. 6.8 Contribution to the graviton 2-point function from matter loops

We start with the tadpole correction, using the dimensional regularization technique, which enables us to capture the running of the parameters of the theory.

Tadpole—At one-loop, the scalar field contributes to the graviton tadpole through the 3-vertex $\hat{v}^{ab}\partial_a\psi\partial_b\psi$ represented in Fig. 6.7. Explicit calculation of this vertex gives

$$\mathcal{A}^{(1pt)} = \int \frac{d^4k}{(2\pi)^4}\frac{\hat{v}^{ab}k_ak_b}{k^2+M^2} = \frac{1}{4}M^4[\hat{v}]J_{M,1}, \tag{6.68}$$

where

$$J_{M,1} = \frac{1}{M^4}\int \frac{d^4k}{(2\pi)^4}\frac{k^2}{k^2+M^2}, \tag{6.69}$$

as explained in Appendix A.3.

We quote for completeness the individual corrections from each Feynman diagram corresponding to a given n-point function.

2-point function—There are two Feynman diagrams which contribute to the corrected 2-point function, which arise respectively from the cubic and quartic interactions in the action (6.67).

Evaluation of these one-loop diagrams shown in Fig. 6.8 gives

$$\mathcal{A}^{(2pt)}_{(a)} = 2\hat{v}^{ab}\hat{v}^{cd}\int \frac{d^4k}{(2\pi)^4}\frac{k_ak_bk_ck_d}{(k^2+M^2)^2} = \frac{1}{4}M^4\left(2[\hat{v}^2]+[\hat{v}]^2\right)J_{M,1},$$
$$\mathcal{A}^{(2pt)}_{(b)} = -3\int \frac{d^4k}{(2\pi)^4}\frac{(\hat{v}^2)^{ab}k_ak_b}{(k^2+M^2)} = -\frac{3}{4}M^4\,[\hat{v}^2]\,J_{M,1},$$

so that the total contribution to the 2-point function is

$$\mathcal{A}^{(2pt)} = \mathcal{A}^{(2pt)}_{(a)} + \mathcal{A}^{(2pt)}_{(b)} = \frac{1}{4}M^4([\hat{v}]^2-[\hat{v}^2])J_{M,1}. \tag{6.70}$$

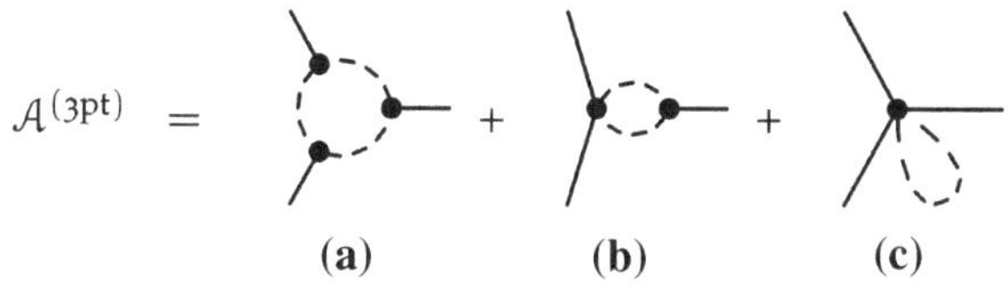

Fig. 6.9 One-loop contributions to the 3-point function

3-point function—The three-point scattering amplitude will receive corrections from the diagrams depicted in Fig. 6.9 which give the following contributions

$$\begin{aligned}\mathcal{A}^{(3pt)}_{(a)} &= \frac{M^4}{4}\left([\hat{v}]^3 + 6[\hat{v}][\hat{v}^2] + 8[\hat{v}^3]\right) J_{M,1},\\ \mathcal{A}^{(3pt)}_{(b)} &= -\frac{9M^4}{4}\left([\hat{v}^2][\hat{v}] + 2[\hat{v}^3]\right) J_{M,1},\\ \mathcal{A}^{(3pt)}_{(c)} &= 3M^4[\hat{v}^3]\, J_{M,1}.\end{aligned}$$

We conclude the total 3-point function goes as

$$\mathcal{A}^{(3pt)} = \frac{M^4}{4}\left(2[\hat{v}^3] + [\hat{v}]^3 - 3[\hat{v}][\hat{v}^2]\right) J_{M,1}. \tag{6.71}$$

4-point function—The Feynman diagrams contributing to the corrected 4-point function are those in Fig. 6.10 and they give the following contributions

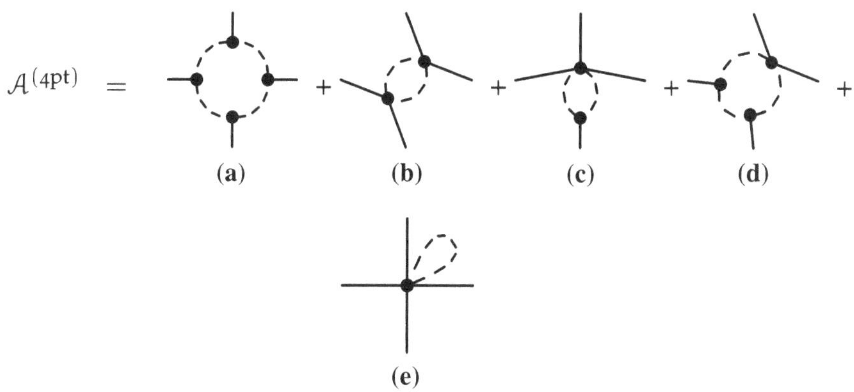

Fig. 6.10 One-loop contributions to the 4-point function

$$\begin{aligned}
\mathcal{A}^{(4\mathrm{pt})}_{(a)} &= \mathrm{M}^4\Big(12[\hat{\mathrm{v}}^4] + 8[\hat{\mathrm{v}}][\hat{\mathrm{v}}^3] + 3[\hat{\mathrm{v}}^2]^2 + 3[\hat{\mathrm{v}}^2][\hat{\mathrm{v}}]^2 + \frac{1}{4}[\hat{\mathrm{v}}]^4\Big)\mathrm{J}_{\mathrm{M},1},\\
\mathcal{A}^{(4\mathrm{pt})}_{(b)} &= \frac{27\mathrm{M}^4}{4}\Big([\hat{\mathrm{v}}^2]^2 + 2[\hat{\mathrm{v}}^4]\Big)\,\mathrm{J}_{\mathrm{M},1},\\
\mathcal{A}^{(4\mathrm{pt})}_{(c)} &= 12\mathrm{M}^4\Big([\hat{\mathrm{v}}^3][\hat{\mathrm{v}}] + 2[\hat{\mathrm{v}}^4]\Big)\,\mathrm{J}_{\mathrm{M},1},\\
\mathcal{A}^{(4\mathrm{pt})}_{(d)} &= -\frac{9\mathrm{M}^4}{2}\Big([\hat{\mathrm{v}}^2][\hat{\mathrm{v}}]^2 + 2[\hat{\mathrm{v}}^2]^2 + 2[\hat{\mathrm{v}}][\hat{\mathrm{v}}^3] + 8[\hat{\mathrm{v}}^4] + 2[\hat{\mathrm{v}}][\hat{\mathrm{v}}^3]\Big)\,\mathrm{J}_{\mathrm{M},1},\\
\mathcal{A}^{(4\mathrm{pt})}_{(e)} &= -15\mathrm{M}^4[\hat{\mathrm{v}}^4]\,\mathrm{J}_{\mathrm{M},1},
\end{aligned}$$

so that the total 4-point function is given by

$$\mathcal{A}^{(4\mathrm{pt})} = \frac{\mathrm{M}^4}{4}\Big([\hat{\mathrm{v}}]^4 - 6[\hat{\mathrm{v}}^4] - 6[\hat{\mathrm{v}}^2][\hat{\mathrm{v}}]^2 + 3[\hat{\mathrm{v}}^2]^2 + 8[\hat{\mathrm{v}}][\hat{\mathrm{v}}^3]\Big)\,\mathrm{J}_{\mathrm{M},1}. \tag{6.72}$$

Equations (6.70)–(6.72) have the precise coefficients to produce a running of the cosmological constant, as shown in Eq. (6.73).

Up to the 4-point function and using Eqs. (6.68), (6.70)–(6.72), the counterterms which ought to be added to the original action (6.67) organize themselves into

$$\begin{aligned}
\mathcal{L}_{\mathrm{CT}} &= -\Big(\mathcal{A}^{(1\mathrm{pt})} + \frac{1}{2!}\mathcal{A}^{(2\mathrm{pt})} + \frac{1}{3!}\mathcal{A}^{(3\mathrm{pt})} + \frac{1}{4!}\mathcal{A}^{(4\mathrm{pt})}\Big)\\
&= -\frac{\mathrm{M}^4}{4}\Big([\hat{\mathrm{v}}] + \frac{1}{2!}([\hat{\mathrm{v}}]^2 - [\hat{\mathrm{v}}^2]) + \frac{1}{3!}([\hat{\mathrm{v}}]^3 + 2[\hat{\mathrm{v}}^3] - 3[\hat{\mathrm{v}}][\hat{\mathrm{v}}^2])\\
&\quad + \frac{1}{4!}([\hat{\mathrm{v}}]^4 - 6[\hat{\mathrm{v}}^4] - 6[\hat{\mathrm{v}}]^2[\hat{\mathrm{v}}^2] + 3[\hat{\mathrm{v}}^2]^2 + 8[\hat{\mathrm{v}}][\hat{\mathrm{v}}^3])\Big)\mathrm{J}_{\mathrm{M},1}\\
&= -\frac{\mathrm{M}^4}{4}\sqrt{g}\,\mathrm{J}_{\mathrm{M},1}.
\end{aligned} \tag{6.73}$$

Note that the last line is only technically correct if we include the zero-point function, which we can do (it is a vacuum bubble). We conclude that the matter loops renormalize the cosmological constant, which is the only potential term one can obtain from integrating out matter loops, in agreement with the findings of 't Hooft and Veltman (1974), Park (2011). Importantly, matter loops do not affect the structure of the graviton potential.

Higher n-point functions—From the one-loop effective action argument, we know that all the n-point functions will receive contributions which will eventually repackage into the normalization of the cosmological constant. Seeing this explicitly at the perturbative level is nevertheless far less trivial, but we give a heuristic argument here. Taking the metric defined in Eq. (6.45), the expansion of the determinant of the metric to quartic order in $\hat{\mathrm{v}}$ as given in (6.73) is, in fact, *exact*. The finite nature of the running of the cosmological constant in (6.73) is therefore no accident.

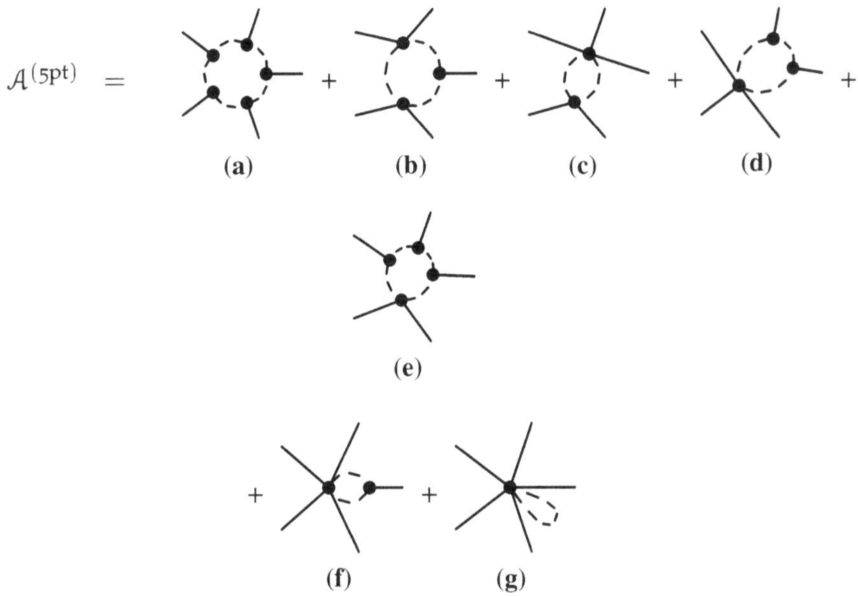

Fig. 6.11 One-loop contributions to the 5-point function

To show we would arrive at the same conclusion by explicit computation, consider the one-loop correction to the 5-point function. In this case, there are five Feynman diagrams which contribute at the same order for the quantum corrections, depicted in Fig. 6.11. From the interactions in the Euclidean action (6.67), we find

$$
\begin{aligned}
\mathcal{A}^{(5pt)} &= \frac{M^4}{4}\bigg([\hat{v}]^5 + 6[\hat{v}^5] - \frac{15}{2}[\hat{v}][\hat{v}^4] + 5[\hat{v}^3][\hat{v}]^2 \\
&\quad -5[\hat{v}^3][\hat{v}^2] + \frac{15}{4}[\hat{v}][\hat{v}^2]^2 - \frac{5}{2}[\hat{v}^2][\hat{v}]^3\bigg) J_{M,1} \\
&\equiv 0,
\end{aligned}
$$

which vanishes identically in four dimensions, as noted in de Rham and Gabadadze (2010). We can proceed in a similar manner to show that the same will be true for all the n-point functions, with $n > 5$. This supports the consistency of the formalism introduced in (6.45) and explicitly agrees with the findings for GR as well as with the direct computation of the one-loop effective action.

Having shown the quantum stability of the massive gravity potential at one-loop, one can see that the same remains true for any number of loops provided there are no virtual gravitons running in the internal lines.

6.2.3 Quantum Corrections from Graviton Loops in the Vielbein Language

In the previous section we have studied in detail the quantum corrections to the potential for massive gravity arising from matter running in loops. We concluded that these quantum corrections could be resummed and interpreted as the renormalization of the cosmological constant. Therefore, we have shown that such loops are completely harmless to the special structure of the ghost-free interaction potential.

Now we push this analysis forward by studying quantum corrections from graviton loops. We start by considering one-loop diagrams, and since we are interested in the IR limit of the theory, we set the external momenta to zero, as before. We will again focus on the running of the interaction couplings, and thus apply dimensional regularization.

Based on studies within the decoupling limit (de Rham et al. 2012), we expect the quantum corrections to the graviton mass to scale as $\delta m^2 \sim m^4/M_{Pl}^2 \sim 10^{-120}\, m^2$. Even though such corrections are parametrically small, a potential problem arises if they detune the structure of the interaction potential. If this happens, ghosts arising at a scale much smaller than the Planck mass could in general plague the theory, rendering it unstable against quantum corrections. To show how such corrections could arise, we organise the loop diagrams in powers of the free parameters $\tilde{\alpha}_3$ and $\tilde{\alpha}_4$ of Eq. (6.50).

To draw a comparison to the previous computation in the metric formulation, the potential interactions in Eqs. (6.51)–(6.53) are now finite in the fluctuations v whereas in the metric formulation they were not. We start by studying the quantum corrections arising at the linear order in the potential parameters $\tilde{\alpha}_{2,3,4}$. Since $\tilde{\mathcal{U}}_2$ is precisely quadratic in v, it leads to no corrections. Next we focus on $\tilde{\mathcal{U}}_3$ in Eq. (6.50), which is cubic in v and therefore can in principle renormalize the tadpole at one-loop. The tadpole contribution yields

$$\mathcal{A}^{(1pt,3vt)} = -\frac{5}{8}\tilde{\alpha}_3 \frac{M^4}{M_{Pl}}[v]J_{M,1}, \tag{6.74}$$

which on its own is harmless (this would correspond to the potential $\tilde{\mathcal{U}}_1$ which we have not included in (6.50), but which is also ghost-free (Hassan and Rosen 2012b)). The last potential term $\tilde{\mathcal{U}}_4$ is quartic in h, as shown in (6.53). This interaction vertex leads to quantum corrections to the 2-point function as shown in Fig. 6.12,

Fig. 6.12 One-loop contribution to the 2-point correlation function from a graviton internal line

$$\mathcal{A}^{(2pt,4vt)} = \mathcal{A}^{(2,4)}_{abcd}\, h^{ab} h^{cd} = 5\tilde{\alpha}_4 \frac{m^4}{M_{Pl}^2} \left([v^2] - [v]^2\right) J_{M,1} \;\; \propto \;\; \tilde{\mathcal{U}}_2(v), \quad (6.75)$$

where we have applied dimensional regularization with $J_{M,1}$ given in Eq. (A.19). This is nothing else but the Fierz–Pauli structure, which is ghost-free by construction (Fierz and Pauli 1939)[4] (Fig. 6.12).

Quadratic and other higher order corrections in $\tilde{\alpha}_3$ and $\tilde{\alpha}_4$ on the other hand are less trivial.

We will see however, that this optimistic result will not prevail for other corrections, which will induce the detuning of the interaction potential structure in Eqs. (6.51)–(6.53). To see this we turn to the interactions coming from the Einstein–Hilbert term. Given (6.45) we can write the Einstein–Hilbert term as

$$-\frac{M_{Pl}^2}{2}\sqrt{g}R = v^{\alpha\beta}\hat{\mathcal{E}}^{\mu\nu}{}_{\alpha\beta}\, v_{\mu\nu} + \frac{1}{M_{Pl}} v(\partial v)^2 + \frac{1}{M_{Pl}^2} v^2(\partial v)^2 + \cdots, \quad (6.76)$$

where $\hat{\mathcal{E}}^{\mu\nu}{}_{\alpha\beta}$ is the usual Lichnerowicz operator written explicitly in Eq. (1.27).

Contrary to the potential in (6.51)–(6.53), Eq. (6.76) contains an infinite numbers of interactions in v. The second order Einstein Hilbert action has the following simple form

$$\mathcal{L}^{2nd}_{EH} = \frac{1}{2}\left(v^a{}_a{}^{,b}(v^u{}_{u,b} - 2v_b^u{}_{,u}) + (2v_{au,b} - v_{ab,u})v^{ab,u}\right) \quad (6.77)$$

nevertheless does not give rise to any interactions. Taking, for example, the third

$$\begin{aligned}\mathcal{L}^{3rd}_{EH} = \frac{1}{2}\Big(& v^{ab}(2v^{uw}{}_{,a} v_{uw,b} + v^{w}{}_{w,b}(-2v^{u}{}_{u,a} + 3v^{u}_{a,u}) + 3v^{u}_{a,b} v^{w}{}_{w;u} \\ & + 2v_{ab,u}(-v^{w}{}_{w,u} + v^{w}_{u}{}_{,w}) - 2(3v_{uw,b} + v_{bw,u} - v_{bu,w})v^{u,w}_{a}) \\ & + v^{a}{}_{a}(v^{b,u}_{b}(v^{w}{}_{w,u} - 2v^{w}_{u}{}_{,w}) + (2v_{bw,u} - v_{bu,w})v^{bu,w})\Big) \end{aligned} \quad (6.78)$$

and quartic order interactions from the Einstein–Hilbert term,

$$\begin{aligned}\mathcal{L}^{4th}_{EH} = \frac{1}{4}\Big(& v^{ab}(v^{uw}(v^{c}_{a}{}_{;u}(9v_{wc;b} - 8v_{bc,w}) + v_{au}{}^{,c}(-v_{bw,c} + 10v_{bc,w}) \\ & + 2(-4v_{au,b}v^{c}{}_{c,w} + v_{ab,u}(-3v^{c}_{w}{}_{c} + 4v^{c}{}_{c,w}) + v_{ab}{}^{,c}(v_{uw,c} - 3v_{uc,w}))) \\ & + v^{u}_{a}(-6v^{wc}{}_{,b}v_{wc,u} + 2v^{c}{}_{c,u}(3v^{w}{}_{w,b} - 4v^{w}_{b}{}_{,w}) \\ & + v^{w,c}_{b}(-5v_{uw,c} + 16v_{wc,u} + 5v_{uc,w}) \\ & - 4v^{c}_{w}{}_{,c}v^{,w}_{bu} + 4v^{c}{}_{c,w}(-2v^{w}_{b}{}_{,u} + v^{,w}_{bu})) + v_{ab}(v^{uw,c}(v_{uw,c} - 2v_{uc,w}) \\ & - (-2v_{w}{}^{c}{}_{,c} + v^{c}{}_{c,w})v^{u}{}_{u}{}^{,w})) + v^{a}{}_{a}(2v^{bu}(2v^{wc}{}_{;b}v_{wc,u} + 3v_{b}{}^{w}{}_{,u}v^{c}{}_{c,w} \end{aligned}$$

[4]At the quadratic level, the Fierz–Pauli term is undistinguishable from the ghost-free potential term $\tilde{\mathcal{U}}_2$

$$
\begin{aligned}
&+ v^{c}{}_{c,u}(-2v^{w}{}_{w,b} + 3v_{b}{}^{w}{}_{,w}) \\
&- 2v_{b}{}^{w,c}(-v_{uw,c} + 3v_{wc,u} + v_{uc,w}) + 2(v_{w}{}^{c}{}_{,c} - v^{c}{}_{c,w})v_{bu}^{;w}) \\
&+ v^{b}{}_{b}(v^{uw,c}(-v_{uw,c} + 2v_{uc,w}) + (-2v_{w}{}^{c}{}_{,c} + v^{c}{}_{c,w})v^{u}{}_{u}{}^{,w}))\Big)
\end{aligned}
\tag{6.79}
$$

we find they do not generate any radiative correction to the tadpole

$$
\mathcal{A}_{\mathrm{EH}}^{(1\mathrm{pt},3\mathrm{vt})} = 0, \tag{6.80}
$$

but they do contribute to the 2-point function as follows

$$
\mathcal{A}_{\mathrm{EH}}^{(2\mathrm{pt})} = \frac{35}{12}\frac{M^4}{M_{\mathrm{Pl}}^2}(4[v^2] - [v]^2)J_{M,1}. \tag{6.81}
$$

The result above also does not preserve the Fierz–Pauli structure and is thus potentially dangerous.

6.2.3.1 Detuning of the Potential Structure

From the above we conclude that quantum corrections from graviton loops can spoil the structure of the ghost-free potential of massive gravity required at the classical level to avoid propagating ghosts. Interestingly, this detuning does not arise from the potential interactions at leading order in the parameters $\tilde{\alpha}_{3,4}$ but does arise from the kinetic Einstein–Hilbert term. Symbolically, the detuning of the potential occurs at the scale

$$
\mathcal{L}_{\mathrm{qc}} \sim \frac{m^4}{M_{\mathrm{Pl}}^n}\, v^n, \tag{6.82}
$$

where m is the graviton mass. When working around a given background for $v = \bar{v}$ (which can include the helicity-0 mode, π), this leads to a contribution at quadratic order which does *not* satisfy the Fierz–Pauli structure,

$$
\mathcal{L}_{\mathrm{qc},\,\bar{v}} \sim \frac{m^4 \bar{v}^{n-2}}{M_{\mathrm{Pl}}^n}\, v^2. \tag{6.83}
$$

Reintroducing the canonically normalized helicity-0 mode as $v_{\mu\nu} = \partial_\mu \partial_\nu \pi / m^2$, this implies a ghost for the helicity-0 mode

$$
\mathcal{L}_{\mathrm{qc},\,\bar{v}} \sim \frac{\bar{v}^{n-2}}{M_{\mathrm{Pl}}^n}\,(\partial^2\pi)^2 \sim \frac{1}{m_{\mathrm{ghost}}^2}\,(\partial^2\pi)^2 \qquad \text{with} \qquad m_{\mathrm{ghost}} = \left(\frac{M_{\mathrm{Pl}}}{\bar{v}}\right)^{n/2}\bar{v}. \tag{6.84}
$$

For interactions with $n \geq 3$, the mass of the ghost, m_{ghost}, can be made arbitrarily *small* by switching on an arbitrarily *large* background configuration for $\bar{v}$. This is clearly a problem since *large* backgrounds ($\bar{v} \gtrsim M_{Pl}$, or alternatively $\partial^2\bar{\pi} \gtrsim \Lambda_3^3$) are important for the Vainshtein mechanism (1972) to work and yet they can spoil the stability of the theory.

6.2.3.2 One-Loop Effective Action

We have shown that quantum corrections originated both from the potential as well as from the Einstein–Hilbert term in general destabilize the ghost-free interactions of massive gravity. This happens in a way which cannot be accounted for by either a renormalization of the coefficients of the ghost-free mass terms, or by a cosmological constant. This detuning leads to a ghost whose mass can be made arbitrarily small if there is a sufficiently large background source. From the decoupling limit analysis, we know that it is always possible to make the background source large enough without going outside of the regime of the effective field theory since we have

$$\bar{v} \sim \frac{1}{m^2}\partial\partial\bar{\pi} = M_{Pl}\frac{1}{\Lambda_3^3}\partial\partial\bar{\pi}\,. \tag{6.85}$$

As an effective field theory we are allowed to make $\partial\partial\pi \gg \Lambda_3^3$ provided $\partial^3\pi/(1+\partial^2\pi/\Lambda_3^3) \ll \Lambda_3^4$. In other words, as long as derivatives of the background $\bar{v}$ are small $\partial \ll \Lambda$, the magnitude of the background may be large $\bar{v} \gg M_{Pl}$.

The resolution of this problem in this case is that one also needs to take into account the redressing of the operators in the interaction potential. In this section we will investigate how the Vainshtein mechanism operates in protecting the effective action from the appearance of dangerous ghosts below the Planck scale.

Our previous approach involved explicit calculation of loop diagrams to evaluate the quantum corrections to the massive gravity potential. Here, we shall focus on the formalism of the one-loop effective action to confirm the destabilization result and provide a more generic argument. Since the quadratic potential $\tilde{\mathcal{U}}_2$ in (6.51) has no non-linear interactions, we can take it as our sole potential term and consider all the graviton self-interactions arising from the Einstein–Hilbert term in Eq. (6.76). For simplicity, and without loss of generality, we therefore consider in what follows the specific theory of massive gravity

$$\mathcal{L} = \frac{M_{Pl}^2}{2}\sqrt{g}R - \frac{1}{4}M_{Pl}^2 m^2 \tilde{\mathcal{U}}_2[v]. \tag{6.86}$$

We start by splitting the field $v_{\mu\nu}$ into a constant background $\bar{v}_{\mu\nu}$ and a perturbation $\delta v_{\mu\nu}(x)$ which, in the language of the previous sections, will be the field running in the loops. We thus write $v_{\mu\nu}(x) = \bar{v}_{\mu\nu} + \delta v_{\mu\nu}(x)$. Up to quadratic order in the perturbation δv

$$\begin{aligned}\mathcal{L} &= \frac{1}{2}\delta v_{\alpha\beta}\left(\hat{\mathcal{E}}^{\mu\nu\alpha\beta} + m^2\left(\delta^{\mu\alpha}\delta^{\nu\beta} - \delta^{\mu\nu}\delta^{\alpha\beta}\right)\right)\delta v_{\mu\nu} \\ &\quad + \left(\frac{1}{M_{Pl}}\bar{v} + \frac{1}{M_{Pl}^2}\bar{v}^2 + \cdots\right)(\partial\delta v)^2 \\ &= \frac{1}{2}\delta v_{\alpha\beta}\left(G^{-1\ \mu\nu\alpha\beta} + M^{\mu\nu\alpha\beta}(\bar{v})\right)\delta v^{\mu\nu} \\ &\equiv \frac{1}{2}\delta v_{\alpha\beta}\,\tilde{M}^{\mu\nu\alpha\beta}\,\delta v_{\mu\nu}, \end{aligned} \tag{6.87}$$

where $M^{\mu\nu\alpha\beta}(\bar{v}) = \left(\frac{1}{M_{Pl}}\bar{v} + \frac{1}{M_{Pl}^2}\bar{v}^2 + \cdots\right)\partial^2$ symbolizes all the interactions in the Einstein–Hilbert term. G^{-1} is the inverse of the massive graviton propagator. Following the same analysis of Sect. 6.2.2.1, the one-loop effective action is then given by

$$\begin{aligned}\mathcal{L}_{\text{eff}} &= -\frac{1}{2}\log\det\left(\frac{1}{\mu^2}\left\{G^{-1\ \mu\nu\alpha\beta} + M^{\mu\nu\alpha\beta}(\bar{v})\right\}\right) \\ &\supseteq -\frac{1}{2\mu^2}\left(\int\frac{d^4k}{(2\pi)^4}\frac{f_{\mu\nu\alpha\beta}M^{\mu\nu\alpha\beta}}{(k^2+m^2)}\right. \\ &\quad \left. -\frac{1}{2}\int\frac{d^4k}{(2\pi)^4}\frac{f_{\mu\nu\alpha\beta}M^{\mu\nu ab}f_{abcd}M^{cd\alpha\beta}}{\left(k^2+m^2\right)^2} + \cdots\right) \end{aligned} \tag{6.88}$$

where $M^{\mu\nu\alpha\beta}(\bar{v})$ is expanded in Fourier space and depends explicitly on a derivative structure (and so, on the momentum k). Here μ denotes again a renormalization scale, which ought to be introduced as a consequence of the renormalization procedure, and to preserve the dimensional analysis. Equation (6.88) sources an effective potential which goes as $m^4\mathcal{F}(\bar{v}/M_{Pl})J_{1,m}$ where $\mathcal{F}$ denotes an infinite series in powers of $\bar{v}$. This result implies a running of the effective potential.

To gain some insight on the form of this effective potential, we focus on the specific case of a conformally flat background where $\bar{v}_{\mu\nu} = \lambda\,\delta_{\mu\nu}$, for some real-valued λ. It follows[5]

$$M^{\mu\nu\alpha\beta}(\bar{v}_{\mu\nu} = \lambda\,\delta_{\mu\nu}) = 0, \tag{6.90}$$

---

[5]This can be seen more explicitly, by writing the operator M in terms of the background metric $\bar{\gamma}_{ab}$, recalling that $\bar{g}_{ab} = \bar{\gamma}_{ac}\bar{\gamma}_{bd}\delta^{cd}$ and the metric g_{ab} is given in terms of $\bar{\gamma}_{ab}$ and the field fluctuation v_{ab} as in (6.44). Then it follows that symbolically,

$$M^{abcd} \sim (\delta^a_\mu\bar{\gamma}_{\nu\rho}\delta^{\rho b})(\delta^c_\alpha\bar{\gamma}_{\beta\sigma}\delta^{\sigma d})(\sqrt{\bar{g}}\,\bar{g}^{\mu\nu}\bar{g}^{\alpha\beta}\bar{g}^{\gamma\delta}\partial_\gamma\partial_\delta), \tag{6.89}$$

where the two first terms in bracket arise from the transition to the 'vielbein-inspired' metric fluctuation and the last term is what would have been otherwise the standard linearized Einstein–Hilbert term on a constant background metric $\bar{g}_{ab}$. Written in this form, M is manifestly conformally invariant.

which means that in this case all interactions are lost and (6.88) also vanishes. This implies that the effective potential for a generic $\bar{v}_{\mu\nu}$ has to be of the form

$$\mathcal{L}_{\rm eff} = c_1 \left([\bar{v}]^2 - 4[\bar{v}^2]\right) + \left(c_2[\bar{v}]^3 + c_3[\bar{v}^2][\bar{v}] - (16c_2 + 4c_3)[\bar{v}^3]\right) + \cdots , \qquad (6.91)$$

for some coefficients c_1, c_2 and c_3. The explicit form of these coefficients can be read off by computing specific Feynman diagrams corresponding to the Einstein–Hilbert interactions, or by considering a more general background metric $\bar{v}_{\mu\nu}$. For instance, c_1 corresponds to the coefficient in Eq. (6.81), $c_1 = \frac{35}{12}\frac{M^4}{M_{\rm Pl}^2}$. It is apparent that this structure is very different from that of the ghost-free potential of Eqs. (6.51)–(6.53). This confirms the results obtained in the previous sections.

At what scale does this running arise? Let us first concentrate on the quadratic term in (6.91). Since the helicity-0 mode π enters as $v_{\mu\nu} = \partial_\mu \partial_\nu \pi / m^2$, that term would lead to a correction of the form

$$\mathcal{L}^{(2)}_{\rm eff} = \frac{m^4}{m_{\rm Pl}^2}\left([\bar{v}]^2 - 4[\bar{v}^2]\right) J_{m,1} \sim \frac{1}{M_{\rm Pl}^2}(\Box\pi)^2 \ln(M^2/\mu^2). \qquad (6.92)$$

This would excite a ghost at the Planck scale. Hence this contribution on its own is harmless. Next we consider the effect of the cubic interactions,

$$\mathcal{L}^{(3)}_{\rm eff} = \frac{m^4}{M_{\rm Pl}^3}[\bar{v}]^3 J_1 \sim \frac{1}{M_{\rm Pl}^3 m^2}(\Box\pi)^3 \ln(m^2/\mu^2). \qquad (6.93)$$

We now elaborate on the general argument mentioned in Sect. 6.2.3.1. We take a background configuration for π which is above the scale $\Lambda_3 = (M_{\rm Pl} m^2)^{1/3}$ for the Vainshtein mechanism to work. This will induce a splitting of the helicity-0 mode $\partial^2\pi = \partial^2\bar{\pi} + \partial^2\delta\pi$, with $\partial^2\bar{\pi} \sim M_{\rm Pl} m^2/\kappa$ and $\kappa < 1$. Then the operator in (6.93) could lead to a ghost at a scale much lower than the Planck scale,

$$\mathcal{L}^{(3)}_{\rm eff} \sim \frac{1}{M_{\rm Pl}^2\,\kappa}(\Box\pi)^2 \ln(m^2/\mu^2). \qquad (6.94)$$

Thus by turning on a *large* background, thereby making κ *smaller*, the scale at which the ghost arises becomes smaller and smaller, and eventually comparable to Λ_3 itself. This renders the theory unstable, as argued in Sect. 6.2.3.1. However, by assuming a large background we also need to understand its effect on the original operators via the Vainshtein mechanism (Vainshtein 1972).

6.2.3.3 Vainshtein Mechanism at the Level of the One-Loop 1PI

The formalism of the one-loop effective action makes the Vainshtein mechanism particularly transparent as far as the redressing of the interaction potential is concerned. We further split the field $\bar{v}$ into a large background configuration $\bar{\bar{v}}_{\mu\nu}$ and a perturbation $\tilde{v}_{\mu\nu} \sim \partial_\mu \partial_\nu \pi / M^2$, such that $\bar{v}_{\mu\nu} = \bar{\bar{v}}_{\mu\nu} + \tilde{v}_{\mu\nu}$. Since $\bar{\bar{v}}_{\mu\nu}$ satisfies the equations of motion we have $\tilde{M}'(\bar{v})|_{\bar{\bar{v}}} = 0$, with $\tilde{M}$ defined as in Eq. (6.87). We proceed as before and expand the one-loop effective action up to second order in $\tilde{v}_{\mu\nu}$, as follows

$$\begin{aligned}\mathcal{L}_{\rm eff} &= -\frac{1}{2}{\rm Tr}\log\left(\frac{1}{\mu^2}\tilde{M}^{\mu\nu\alpha\beta}(\bar{v})\right)\\ &\supseteq -\frac{1}{2}{\rm Tr}\,\frac{G_{\mu\nu\alpha\beta}\tilde{M}''^{\mu\nu\alpha\beta}(\bar{\bar{v}})}{G_{\mu\nu\alpha\beta}\tilde{M}^{\mu\nu\alpha\beta}(\bar{\bar{v}})}\,\tilde{v}^2\\ &\sim\frac{m^4}{M_{\rm Pl}^2}\frac{\tilde{M}''(\bar{\bar{v}})}{\tilde{M}(\bar{\bar{v}})}[\tilde{v}]^2 \sim \frac{\Xi}{M_{\rm Pl}^2}\,(\Box\pi)^2,\end{aligned} \tag{6.95}$$

where the last line is symbolic,[6] and for simplicity of notation we have denoted the combination $\Xi \equiv \tilde{M}''(\bar{\bar{v}})/\tilde{M}(\bar{\bar{v}})$. It follows that the mass of the ghost is $M^2_{\rm ghost} = \Xi^{-1}M^2_{\rm Pl}$. Provided we can show that $\Xi \lesssim 1$, the ghost will arise at least at the Planck scale and the theory will always be under control.

Redressing the one-loop effective action—The basis of the argument goes as follows. As explained previously, the Vainshtein mechanism relies on the fact that the background configuration can be large, and thus $\tilde{M}''$ can in principle be large, which in turn can make Ξ large and lower the mass of the ghost. However, as we shall see in what follows, configurations with large M'' automatically lead to large M as well. This implies that Ξ is always bounded $\Xi \lesssim 1$, and the mass of the ghost induced by the detuning/destabilization of the potential from quantum corrections is always at the Planck scale or beyond.[7]

Computing the mass of the ghost—To make contact with an explicit calculation we choose, for convenience, a background configuration for the metric which is spacetime independent, in particular the following background metric $\bar{g}_{ab} = {\rm diag}\{\lambda_0^2, \lambda_1^2, \lambda_1^2, \lambda_1^2\} = \bar{\gamma}^2_{ab}$, and compute the possible combinations appearing in Ξ. We define

$$\frac{\partial^2\tilde{M}/\partial\bar{\bar{v}}_{\alpha\beta}\partial\bar{\bar{v}}_{\gamma\delta}}{\tilde{M}} \equiv \Xi_{(\alpha\beta,\gamma\delta)}, \tag{6.96}$$

and use units for which $M^2_{\rm Pl} = 1$.

---

[6]In particular, by dimensional analysis, one should think of the schematic form for the effective Lagrangian as containing a factor of $1/\mu^2$, where μ carries units of [mass].

[7]The only way to prevent Ξ from being $\lesssim 1$ is to consider a region of space where some eigenvalues of the metric itself vanish, which would be for instance the case at the horizon of a black hole. However as explained in Deffayet and Jacobson (2012), Koyama et al. (2011a, b), Berezhiani et al. (2012), in massive gravity these are no longer coordinate singularities, but rather real singularities. In massive gravity, black hole solutions ought to be expressed in such a way that the eigenvalues of the metric never reach zero apart at the singularity itself. Thus we do not need to worry about such configurations here (which would correspond to $\lambda = 1$ in what follows).

In general, the components of $\tilde{M}$ can be split into three categories. First, some components do not depend on the background, and are thus explicitly *independent* of λ_0 and λ_1. In this case $\tilde{M}'' = \Xi = 0$ trivially. Second, other components are of the form

$$\tilde{M} \sim \frac{\lambda_0}{\lambda_1}\left(k_i k_j + M^2\right), \tag{6.97}$$

where i and j are spatial indices. In this case $\Xi_{(00,00)} = 0$, whereas $\Xi_{(ii,ii)} \sim \lambda_1^{-2} \lesssim 1$.

Finally the remaining category contains terms which only depend on some power of the ratio of the components of the background metric, λ_0/λ_1 or λ_1/λ_0. The structure of the components of $\tilde{M}$ in this case is of the form

$$\tilde{M} \sim \left(\frac{\lambda_1}{\lambda_0}\right)^p k_0^2 + \left(\frac{\lambda_0}{\lambda_1}\right)^q \sum_i b_i k_i^2 + \frac{\lambda_0}{\lambda_1} M^2, \tag{6.98}$$

in which i denotes a spatial component, b_i, p and q are integer numbers such that $p, q \geq 1$. The last term represents the (non-vanishing) structure of the mass term.[8] Let us work out the possible second derivatives in detail. On the one hand

$$\Xi_{(00,00)} \sim \frac{\left(\frac{\lambda_1}{\lambda_0}\right)^p k_0^2 \lambda_0^{-2} + \left(\frac{\lambda_0}{\lambda_1}\right)^q \lambda_0^{-2} \sum_i b_i k_i^2}{\left(\frac{\lambda_1}{\lambda_0}\right)^p k_0^2 + \left(\frac{\lambda_0}{\lambda_1}\right)^p \frac{\lambda_0}{\lambda_1} \sum_i b_i k_i^2 + \text{mass term}}, \tag{6.99}$$

where we have ignored factors of order unity to avoid clutter. At first sight this result seems troublesome as it appears to be dependent on the choice of background, and in particular on the hierarchy between λ_0 and λ_1. We will however show that this is *not* the case:

1. if $\lambda_0 \sim \lambda_1 \sim \lambda$, then $\Xi_{(00,00)} \sim \lambda^{-2}$. Since we are interested in incorporating the Vainshtein effect, we shall consider the case when $\lambda \gtrsim 1$, and thus $\Xi \lesssim 1$.
2. if $\lambda_0 \gg \lambda_1 \gtrsim 1$, it follows $\Xi \lesssim \lambda_0^{-2} \lesssim 1$; the same holds true in the case $\lambda_1 \gg \lambda_0 \gtrsim 1$.

On the other hand,

$$\Xi_{(jj,kk)} = \Xi_{(jj,jj)} \sim \frac{\left(\frac{\lambda_1}{\lambda_0}\right)^p k_0^2 \lambda_1^{-2} + \left(\frac{\lambda_0}{\lambda_1}\right)^q \lambda_1^{-2} \sum_i b_i k_i^2 + \frac{\lambda_0}{\lambda_1^3} M^2}{\left(\frac{\lambda_1}{\lambda_0}\right)^p k_0^2 + \left(\frac{\lambda_0}{\lambda_1}\right)^q \sum_i b_i k_i^2 + \frac{\lambda_0}{\lambda_1} M^2}. \tag{6.100}$$

We repeat the previous analysis to show that, regardless of the possible hierarchy between λ_0 and λ_1, the quantum corrections will be parametrically small.

---

[8] The mass term can also arise in the form $(\lambda_1/\lambda_0)\, M^2$ for the components $\tilde{M}^{00ii}$, but the conclusions hereafter remain unchanged.

1. if $\lambda_0 \sim \lambda_1 \sim \lambda$, then $\Xi_{(jj,jj)} \sim \lambda^{-2}$, and hence $\Xi \lesssim 1$.
2. if either $\lambda_0 \gg \lambda_1 \gtrsim 1$ or $\lambda_1 \gg \lambda_0 \gtrsim 1$, then $\Xi \lesssim \lambda_1^{-2} \ll 1$.

Similar conclusions can be drawn for the 'mixed' derivative $\Xi_{(00,ij)}$ or $\Xi_{(0i,0j)}$.

We have shown that, despite appearances, $\Xi \lesssim 1$ independently of the background and without loss of generality. Whenever the Vainshtein mechanism is relevant, that is, when $\lambda_0, \lambda_1 \gtrsim 1$, the redressing of the operators ensures that the mass of the ghost arises at least at the Planck scale. We therefore conclude that at the one-loop level the quantum corrections to the theory described by (6.86) are under control.

6.3 Quantum Corrections in Bimetric Gravity

Another interesting avenue to explore is how the quantum corrections change if we allow for a dynamical reference metric, i.e. in the bimetric theories. Since now the reference metric becomes also dynamical, we have to include the diagrams in which the mixing with the second metric will occur. From our experience and knowledge gained from the quantum corrections in massive gravity, we can make quick estimates about the common and possible different outcomes in bimetric gravity. The results in the following are not published in any paper yet and are therefore preliminary. Let us first recapitulate the important formulas we will need in this section in order to gain some insight into the quantum corrections in bimetric theory. Consider the ghost-free theory of bimetric gravity. Our starting point is the action for bimetric gravity and the matter action sourcing for gravity (Hassan and Rosen 2011),

$$
\begin{aligned}
S = \int d^4x\big[&M_p^2\sqrt{-g}R_g + M_f^2\sqrt{-f}R_f - 2M^4\sqrt{-g}\sum_{n=0}^{4}\beta_n V_n(g^{-1}f) \\
&+\mathcal{L}_m(g, f, \psi_m)\big]
\end{aligned}
\tag{6.101}
$$

where the potential is given by

$$
\sum_{n=0}^{4}\beta_n V_n(g^{-1}f) = \sum_{n=0}^{4}\beta_n e_n(\mathcal{X} = \sqrt{g^{-1}f}) \tag{6.102}
$$

with the elementary symmetric polynomials (de Rham and Gabadadze 2010; de Rham et al. 2011b)

$$
\begin{aligned}
e_0(\mathcal{X}) &= 1 \\
e_1(\mathcal{X}) &= [\mathcal{X}] \\
e_2(\mathcal{X}) &= \frac{1}{2}([\mathcal{X}]^2 - [\mathcal{X}^2])
\end{aligned}
$$

$$e_3(\mathcal{X}) = \frac{1}{6}([\mathcal{X}]^3 - 3[\mathcal{X}][\mathcal{X}^2] + 2[\mathcal{X}^3])$$
$$e_4(\mathcal{X}) = \frac{1}{24}([\mathcal{X}]^4 - 6[\mathcal{X}]^2[\mathcal{X}^2] + 3[\mathcal{X}^2]^2 + 8[\mathcal{X}][\mathcal{X}^3] - 6[\mathcal{X}^4]) \quad (6.103)$$

In difference to the massive gravity theory the potential term is now the potential for both metrics and the forth polynomial will correspond to a potential for $f_{\mu\nu}$ and can not any longer be neglected in the bimetric gravity. It has been shown that this theory of bimetric gravity is free of any ghost-issues (Hassan and Rosen 2012a).

For flat euclidean backgrounds $\bar{g}_{\mu\nu} = \delta_{\mu\nu} = \bar{f}_{\mu\nu}$ we split the mass spectrum of bimetric gravity into massive and massless spin-2 fluctuations (for these flat backgrounds two out of the five β_n parameters are fixed, namely the ones corresponding to a cosmological constant for the $g_{\mu\nu}$ metric and $f_{\mu\nu}$ metric). To be precise, the mass spectrum can be obtained by performing the following metric perturbations (Hassan and Rosen 2012a)

$$g_{\mu\nu} = \delta_{\mu\nu} + \frac{2}{M_{Pl}} h_{\mu\nu} \quad (6.104)$$
$$f_{\mu\nu} = \delta_{\mu\nu} + \frac{2}{M_f} l_{\mu\nu} \quad (6.105)$$

and plugging them back into the action for the bigravity

$$S = \int d^4x \Big\{ - h_{\mu\nu}\hat{\mathcal{E}}^{\mu\nu\alpha\beta} h_{\alpha\beta} - l_{\mu\nu}\hat{\mathcal{E}}^{\mu\nu\alpha\beta} l_{\alpha\beta} - \frac{M^2 M_{eff}^2}{4}\left[(\frac{h^\mu_\nu}{M_{Pl}} - \frac{l^\mu_\nu}{M_f})^2 - (\frac{h^\mu_\mu}{M_{Pl}} - \frac{l^\mu_\mu}{M_f})^2 \right] \Big\} \quad (6.106)$$

where $M_{eff}^2 = (1/M_{Pl}^2 + 1/M_f^2)^{-1}$ is the effective Planck mass. These interactions can then be diagonalized by making the following change of variables

$$\frac{1}{M_{eff}} w_{\mu\nu} = \frac{h_{\mu\nu}}{M_f} + \frac{l_{\mu\nu}}{M_{Pl}}$$
$$\frac{1}{M_{eff}} v_{\mu\nu} = \frac{h_{\mu\nu}}{M_{Pl}} - \frac{l_{\mu\nu}}{M_f} \quad (6.107)$$

such that the action at linear order becomes (Hassan and Rosen 2012a)

$$S = \int d^4x \left\{ -w_{\mu\nu}\hat{\mathcal{E}}^{\mu\nu\alpha\beta} w_{\alpha\beta} - v_{\mu\nu}\hat{\mathcal{E}}^{\mu\nu\alpha\beta} v_{\alpha\beta} - \frac{M^2}{4}\left[[v^2] - [v]^2 \right] \right\}. \quad (6.108)$$

In the unitary gauge $v_{\mu\nu}$ encodes all the five physical degrees of freedom of a massive spin-2 fluctuation (the two helicity-2, the two helicity-1 and the helicity-0 modes), and $w_{\mu\nu}$ encodes the two helicity-2 modes of the massless fluctuation. The Feynman propagator for the massless spin-2 fluctuation w is given by

$$G^{(w)}_{abcd} = \langle w_{ab}(x_1) w_{cd}(x_2)\rangle = f^{(w)}_{abcd} \int \frac{d^4k}{(2\pi)^4} \frac{e^{ik\cdot(x_1-x_2)}}{k^2 - i\varepsilon}, \tag{6.109}$$

where the polarization structure has the usual prefactor of 1/2

$$f^{(w)}_{abcd} = \delta_{a(c}\delta_{bd)} - \frac{1}{2}\delta_{ab}\delta_{cd}. \tag{6.110}$$

Here again $\delta_{a(c}\delta_{bd)} \equiv \frac{1}{2}\delta_{ac}\delta_{bd} + \frac{1}{2}\delta_{ad}\delta_{bc}$. The massive spin-2 field, on the other hand, has the Feynman propagator

$$G^{(v)}_{abcd} = \langle v_{ab}(x_1) v_{cd}(x_2)\rangle = f^{(v)}_{abcd} \int \frac{d^4k}{(2\pi)^4} \frac{e^{ik\cdot(x_1-x_2)}}{k^2 + m^2 - i\varepsilon}, \tag{6.111}$$

with the prefactor of 1/3 in the polarization structure

$$f^{(v)}_{abcd} = \left(\tilde{\delta}_{a(c}\tilde{\delta}_{bd)} - \frac{1}{3}\tilde{\delta}_{ab}\tilde{\delta}_{cd}\right) \quad \text{where} \quad \tilde{\delta}_{ab} = \delta_{ab} + \frac{k_a k_b}{m^2}. \tag{6.112}$$

6.3.1 *Quantum Corrections in the Decoupling Limit of Bimetric Theory*

First of all, it is going to be a trivial statement, that in the decoupling limit of bimetric gravity exactly the same non-renormalization theorem protects the interactions from quantum corrections as it was the case in the massive gravity. The interaction between the two metrics break the two copies of diffeomorphisms down to one, such that in the decoupling limit the interactions are governed by decoupled four helicity-2 modes $h_{\mu\nu}$, $l_{\mu\nu}$, two helicity-1 modes and one helicity-0 π modes giving rise in total to seven propagating helicity modes. The decoupling limit of bigravity represents the limit in which

$$M_{Pl} \to \infty, M_f \to \infty, m \to 0 \quad \text{and} \quad \frac{M_{Pl}}{M_f} = \text{const} \tag{6.113}$$

where M_{Pl} and M_f represent the Planck masses corresponding to the two metrics $g_{\mu\nu}$ and $f_{\mu\nu}$. The resulting limit contains interactions between the two helicity-2 fields and the helicity-0 scalar field π in the following form (Fasiello and Tolley 2013)

$$S = \int d^4x \Big[- h^{\mu\nu} \hat{\mathcal{E}}^{\mu\nu\alpha\beta} h_{\alpha\beta} - l^{\mu\nu} \hat{\mathcal{E}}^{\mu\nu\alpha\beta} l_{\alpha\beta} + \Lambda_3^3 \sum_{n=0}^{4} \left(h^{\mu\nu} X^{(n)}_{\mu\nu} + \frac{M_P}{M_f} l^{\mu\nu} \tilde{Y}^{(n)}_{\mu\nu} \right) \Big] \tag{6.114}$$

where $\hat{\mathcal{E}}$ is the Lichnerowicz operator and the $X_{\mu\nu}$ and $\tilde{Y}_{\mu\nu}$ encode the derivative interactions of the helicity-0 field

$$\begin{aligned} X^{(n)}_{\mu\nu} &= -\frac{1}{2} \frac{\hat{\beta}_n}{(3-n)!n!} \mathcal{E}^{\mu\cdots} \mathcal{E}^{\nu\cdots} (\eta + \Pi)^n \eta^{3-n} \\ \tilde{Y}^{(n)}_{\mu\nu} &= -\frac{1}{2} \frac{\hat{\beta}_n}{(4-n)!(n-1)!} \mathcal{E}^{\mu\cdots} \mathcal{E}^{\nu\cdots} \eta^{(n-1)} (\eta + \Sigma)^n \eta^{4-n} \end{aligned} \tag{6.115}$$

where $\hat{\beta}_n = M_P^2 \beta_n$, $\Pi_{\mu\nu} = \partial_\mu \partial_\nu \pi / \Lambda_3^3$ and $\Sigma_{\mu\nu} = \partial_\mu \partial_\nu \rho / \Lambda_3^3$ and where ρ is the dual description of π via field redefinitions related in a form

$$(\eta + \Sigma) = (\eta + \Pi)^{-1} \tag{6.116}$$

It is a trivial step to convince ourselves that exactly the same argumentation for the non-renormalization theorem used in massive gravity applies here in the bimetric gravity case. The Levi-Civita symbols guaranties that any external particle contracted with any field with or without derivatives in a vertex contributes to a two-derivatives operator acting on this external particle. Therefore, the non-renormalization theorem guaranties that the interactions $h^{\mu\nu} X^{(n)}_{\mu\nu}$ and $l^{\mu\nu} \tilde{Y}^{(n)}_{\mu\nu}$ remain stable in the decoupling limit of bimetric theory.

6.3.2 Beyond the Decoupling Limit

Beyond the decoupling limit, there will be essential differences to the case of massive gravity. The first important difference will be concerning the coupling to the matter fields. There will be essentially three different scenarios which need to be investigated in detail.

The first scenario:—The first case consists of having the matter fields coupled only either to $g_{\mu\nu}$ or to $f_{\mu\nu}$ but never to both of them simultaneously

$$\sqrt{-g} \mathcal{L}_\chi(g, \chi_1) \qquad \text{or} \qquad \sqrt{-f} \mathcal{L}_m(f, \chi_2) \tag{6.117}$$

In this case, at one-loop the matter fields will contribute in form of a cosmological constant for $g_{\mu\nu}$ and $f_{\mu\nu}$. This can be shown in a straightforward way. Consider again for simplicity that the matter fields χ_1 and χ_2 are just massive scalar fields (and perform the change of variables $\chi_1 \to g^{-1/4} \psi$ and $\chi_2 \to f^{-1/4} \phi$ respectively)

$$\mathcal{S}_m(g,\psi) = \int d^4x \left(\frac{1}{2} g^{ab}\partial_a\psi\partial_b\psi + \frac{1}{2}M_\psi^2\psi^2\right)$$
$$\mathcal{S}_m(f,\phi) = \int d^4x \left(\frac{1}{2} f^{ab}\partial_a\phi\,\partial_b\phi + \frac{1}{2}M_\phi^2\,\phi^2\right) \tag{6.118}$$

where we neglect again the terms of the form ∂g and ∂f. Since loops only involve the matter field propagator and are unaware of the propagators of the two metrics, there is no contributions from the bimetric interactions to these quantum corrections, and we recover exactly the same result as in Massive Gravity and General Relativity. This statement becomes trivial when we look at the one loop effective action for the matter loops

$$e^{-S_{1,\mathrm{eff}}^{(\mathrm{matter-loops})}(\bar{g}_{ab},\bar{f}_{ab})} = \int \mathcal{D}\psi e^{-\psi\left(\frac{\delta^2 S}{\delta^2\psi}|_{g_{ab}=\bar{g}_{ab}}\right)\psi-\phi\left(\frac{\delta^2 S}{\delta^2\phi}|_{f_{ab}=\bar{f}_{ab}}\right)\phi}$$
$$= \left(\int \mathcal{D}\psi e^{-\psi\left(\frac{\delta^2 S}{\delta^2\psi}|_{g_{ab}=\bar{g}_{ab}}\right)\psi}\right)\left(\int \mathcal{D}\phi\, e^{-\phi\left(\frac{\delta^2 S}{\delta^2\phi}|_{f_{ab}=\bar{f}_{ab}}\right)\phi}\right) \tag{6.119}$$

with the Coleman–Weinberg effective action given by

$$S_{1,\mathrm{eff}}^{(\mathrm{matter-loops})}(\bar{g}_{ab},\bar{f}_{ab}) = \frac{1}{2}\log\det\left(\frac{\delta^2 S}{\delta^2\psi}\bigg|_{g_{ab}=\bar{g}_{ab}}\right) + \frac{1}{2}\log\det\left(\frac{\delta^2 S}{\delta^2\phi}\bigg|_{f_{ab}=\bar{f}_{ab}}\right)$$
$$= \frac{1}{2}\mathrm{Tr}\log\left(\frac{\delta^2 S}{\delta^2\psi}\bigg|_{g_{ab}=\bar{g}_{ab}}\right) + \frac{1}{2}\mathrm{Tr}\log\left(\frac{\delta^2 S}{\delta^2\phi}\bigg|_{f_{ab}=\bar{f}_{ab}}\right). \tag{6.120}$$

We go now into Fourier space and performe the change of momentum $k_a \to \tilde{k}_a$ such that $g^{ab}k_ak_b = \delta^{ab}\tilde{k}_a\tilde{k}_b$ and so the one loop effective action for the matter field ψ coupled to $g_{\mu\nu}$ yields

$$\mathcal{L}_{1,\mathrm{eff}}^{(\mathrm{matter-loops})}(\bar{g}_{ab}) = \frac{1}{2}\int \frac{d^4k}{(2\pi)^4}\log\left(\frac{1}{2}\bar{g}^{ab}k_ak_b + \frac{1}{2}M_\psi^2\right)$$
$$= \frac{1}{2}\sqrt{\bar{g}}\int \frac{d^4\tilde{k}}{(2\pi)^4}\log\left(\frac{1}{2}\delta^{ab}\tilde{k}_a\tilde{k}_b + \frac{1}{2}M_\psi^2\right)$$
$$= \frac{M_\psi^4}{64\pi^2}\sqrt{\bar{g}}\log(\mu^2), \tag{6.121}$$

where μ is the regularization scale. Similarly, by doing the change of momentum $k_a \to \tilde{k}_a$ with $f^{ab}k_ak_b = \delta^{ab}\tilde{k}_a\tilde{k}_b$ gives for the other metric

$$\mathcal{L}^{(\text{matter-loops})}_{1,\text{eff}}(\bar{f}_{ab}) = \frac{1}{2}\int \frac{d^4k}{(2\pi)^4}\log\left(\frac{1}{2}\bar{f}^{ab}k_a k_b + \frac{1}{2}M_\phi^2\right)$$
$$= \frac{1}{2}\sqrt{\bar{f}}\int \frac{d^4\tilde{k}}{(2\pi)^4}\log\left(\frac{1}{2}\delta^{ab}\tilde{k}_a\tilde{k}_b + \frac{1}{2}M_\phi^2\right)$$
$$= \frac{M_\phi^4}{64\pi^2}\sqrt{\bar{f}}\log(\mu'^2), \tag{6.122}$$

Exactly in the same way as in Sect. 6.2.2.1 these two contributions in the one loop effective action give contributions in form of $\sqrt{g}$ and $\sqrt{f}$. This conclusion can be easily extended to N-metric theories to which the matter fields couple using the one-loop effective action. In an analog way as in massive gravity, we can also perform a perturbative analysis instead of the non-perturbative one-loop effective action, nevertheless one needs to be careful. The real physical degrees of freedom are $w_{\mu\nu}$ and $v_{\mu\nu}$, so these will be the ones running in the perturbative Feynmann diagrams. So one has to couple the matter fields to the metric fluctuations $h_{\mu\nu}$ or $l_{\mu\nu}$

$$\frac{1}{2}\sum_{n\geq 0}(-1)^n\,(n+1)(h^{ab})^n\partial_a\psi\partial_b\psi + \frac{1}{2}M_\psi^2\psi^2$$
$$\frac{1}{2}\sum_{n\geq 0}(-1)^n\,(n+1)(l^{ab})^n\partial_a\,\phi\,\partial_b\,\phi + \frac{1}{2}M_\phi^2\,\phi^2 \tag{6.123}$$

but rephrase them at the Feynmann diagrams level by the corresponding physical degrees of freedom $w_{\mu\nu}$ and $v_{\mu\nu}$, i.e. perform the following replacements

$$h_{\mu\nu} = M_c(M_f v_{\mu\nu} + M_{Pl} w_{\mu\nu})$$
$$l_{\mu\nu} = M_c(-M_{Pl} v_{\mu\nu} + M_f w_{\mu\nu}) \tag{6.124}$$

with $M_c = \frac{M_f M_{Pl}}{M_{eff}(M_f^2+M_{Pl}^2)}$. All these one-loop Feynmann diagrams at the end will sum into a contribution given by the above 1-loop effective action expressions [in a similar way as in massive gravity all the Feynmann diagrams for the n-point functions resumed into a contribution of a cosmological constant (6.73)].

The second scenario:—In the second scenario one should consider the matter fields coupled to both metrics. Are the 1-loop matter contributions still in form of a cosmological constant if we couple the matter fields to the same metrics? We expect that this will very probably yield ghost instabilities. To see this, lets have a look at the one loop effective action and also some explicit calculations of the Feynman diagrams. Now when we do the replacement $\chi \to g^{-1/4}\psi$ we need to be careful since f couples to the same matter field. But we can still perform a similar change of variables but this time with an effective metric $\chi \to (g^{-1/4} + f^{-1/4})\psi$ such that

Fig. 6.13 Tadpole contribution from matter loops with the vertices $h^{ab}\partial_a\partial_b\psi$ and $l^{ab}\partial_a\partial_b\psi$ respectively. The matter field is depicted by the *dashed line*

$$\mathcal{L}_{1,\text{eff}}^{(\text{matter}-\text{loops})}(\bar{g}_{ab},\bar{f}_{ab},)=\frac{1}{2}\int\frac{d^4k}{(2\pi)^4}\log\left(\frac{1}{2}\bar{g}_{\text{eff}}^{ab}k_ak_b+\frac{1}{2}M_\psi^2\right) \tag{6.125}$$

with the effective metric given by

$$g_{\text{eff}}^{ab}=\frac{\sqrt{-g}g^{ab}+\sqrt{-f}f^{ab}}{\sqrt{-g}+\sqrt{-f}} \tag{6.126}$$

Therefore, after the change of momentum $k_a\to\tilde{k}_a$ with $g_{\text{eft}}^{ab}k_ak_b=\delta^{ab}\tilde{k}_a\tilde{k}_b$ this will give rise to a contribution of a cosmological constant for the effective metric $\sqrt{-g_{\text{eff}}}$, but in terms of f and g the new contributions do not give ghost free mass term interactions. To convince ourselves we can compute the contributions perturbatively. We can quickly have a look at the tadpole and two point contributions for the case where the same matter field couples to both metrics. The tadpole contributions are coming from the interactions $h^{ab}\partial_a\partial_b\psi$ and $l^{ab}\partial_a\partial_b\psi$. Computing the tadpole in terms of the physical degrees of freedom give rise to contributions of the form (Fig. 6.13)

$$\begin{aligned}\mathcal{A}^{(1pt)}&=(-1)(-\frac{1}{2})(\frac{1}{4})M_\psi^4M_c\left(\frac{M_f[v]+M_{Pl}[w]}{M_{Pl}}+\frac{-M_{Pl}[v]+M_f[w]}{M_f}\right)J_{M_\psi,1}\\&=\frac{1}{8}M_\psi^4\left(\frac{[h]}{M_{Pl}}+\frac{[l]}{M_f}\right)J_{M_\psi,1}\end{aligned} \tag{6.127}$$

At this point it is worth to mention that if both metrics had the same Planck scales, meaning that the two degrees of freedom $w_{\mu\nu}$ and $v_{\mu\nu}$ were sharing the same scaling then there would be only the tadpole contribution for the massless degree of freedom

$$\mathcal{A}^{(1pt)}|_{M_{Pl}=M_f}=\frac{1}{4}M_\psi^4M_c[w] \tag{6.128}$$

The two point contributions are a little bit more involved but not too difficult to compute. We will first compute the separate diagrams where the metric perturbations do not couple. Let us start with the contributions coming from the matter interactions with h. The interaction $-\frac{1}{2}\frac{h^{ab}}{M_{Pl}}\partial_a\psi\partial_b\psi$ will give rise to a two point function contribution from the diagram with two vertices. This diagram gives rise to a contribution of the following form (Fig. 6.14)

$\mathcal{A}_h^{(2pt,2v)} =$ 

Fig. 6.14 Two point contribution $\mathcal{A}_h^{(2pt,2v)}$ with the vertices $-\frac{1}{2}\frac{h^{ab}}{M_{Pl}}\partial_a\psi\partial_b\psi$

$\mathcal{A}_h^{(2pt,1v)} =$ 

Fig. 6.15 Two point contribution $\mathcal{A}_h^{(2pt,1v)}$ with the vertex $(\frac{3}{4}h_c^a h^{cb} - \frac{1}{4}[h]h^{ab})\partial_a\psi\partial_b\psi$

$$\begin{aligned}\mathcal{A}_h^{(2pt,2v)} &= \frac{1}{2}\frac{1}{4}(2\cdot 2)\frac{1}{24}\frac{M_\psi^4}{M_{Pl}^2}\left([h]^2 + 2[h^2]\right)\cdot(3J_{M_\psi,1}) \\ &= \frac{1}{16}\frac{M_\psi^4}{M_{Pl}^2}\left([h]^2 + 2[h^2]\right)J_{M_\psi,1}\end{aligned} \tag{6.129}$$

On the other hand the interaction to second order in metric perturbation $(\frac{3}{4}h_c^a h^{cb} - \frac{1}{4}h^{ab})\partial_a\psi\partial_b\psi$ will contribute to the tadpole two-point function. This one vertex diagram contributes (Fig. 6.15)

$$\begin{aligned}\mathcal{A}_h^{(2pt,1v)} &= (-1)\frac{M_\psi^4}{M_{Pl}^2}\left(\frac{3}{4}(2)\frac{1}{4}[h^2] - \frac{1}{4}(2)\frac{1}{4}[h]^2\right)J_{M_\psi,1} \\ &= -\frac{1}{8}\frac{M_\psi^4}{M_{Pl}^2}\left(3[h^2] - [h]^2\right)J_{M_\psi,1}\end{aligned} \tag{6.130}$$

Summing the two-point function contributions yields

$$\mathcal{A}_h^{(2pt)} = \mathcal{A}_h^{(2pt,2v)} + \mathcal{A}_h^{(2pt,1v)} = \frac{1}{16}\frac{M_\psi^4}{M_{Pl}^2}(3[h]^2 - 4[h^2]) \tag{6.131}$$

Now we take into account the diagrams with l-perturbations on the external legs. This diagram gives rise to a contribution of the following form (Fig. 6.16)

$\mathcal{A}_l^{(2pt,2v)} =$ 

Fig. 6.16 Two point contribution $\mathcal{A}_l^{(2pt,2v)}$ coming from the interactions of the matter field with the l-perturbations on the vertices

$\mathcal{A}_l^{(2pt,1v)} =$ 

Fig. 6.17 Two point contribution $\mathcal{A}_l^{(2pt,1v)}$ coming from the interactions of the matter field with the l-perturbations on the vertex

$$\mathcal{A}_l^{(2pt,2v)} = \frac{1}{2}\frac{1}{4}(2\cdot 2)\frac{1}{24}\frac{M_\psi^4}{M_{Pl}^2}\left([l]^2+2[l^2]\right)\cdot(3J_{M_\psi,1})$$
$$= \frac{1}{16}\frac{M_\psi^4}{M_{Pl}^2}\left([l]^2+2[l^2]\right)J_{M_\psi,1} \tag{6.132}$$

And the corresponding tadpole two point function This one vertex diagram contributes (Fig. 6.17)

$$\mathcal{A}_l^{(2pt,1v)} = (-1)\frac{M_\psi^4}{M_{Pl}^2}\left(\frac{3}{4}(2)\frac{1}{4}[l^2]-\frac{1}{4}(2)\frac{1}{4}[l]^2\right)J_{M_\psi,1}$$
$$= -\frac{1}{8}\frac{M_\psi^4}{M_{Pl}^2}\left(3[l^2]-[l]^2\right)J_{M_\psi,1} \tag{6.133}$$

Similarly the two Feynmann diagrams for the interactions with the metric perturbation $l_{\mu\nu}$ will sum into

$$\mathcal{A}_l^{(2pt)} = \mathcal{A}_l^{(2pt,2v)} + \mathcal{A}_l^{(2pt,1v)} = \frac{1}{16}\frac{M_\psi^4}{M_{Pl}^2}(3[l]^2-4[l^2]) \tag{6.134}$$

Now last but not least we have to also consider the mixed Feynman diagrams with the different metrics on the external legs. The diagram with the two vertices is basically again constructed by the interactions $-\frac{1}{2}\frac{h^{ab}}{M_{Pl}}\partial_a\psi\partial_b\psi$ and $-\frac{1}{2}\frac{l^{cd}}{M_f}\partial_c\psi\partial_d\psi$. This diagram gives a contribution of the form (Fig. 6.18)

$\mathcal{A}_{hl,1}^{(2pt)} =$ 

Fig. 6.18 Two point contribution $\mathcal{A}_{hl,1}^{(2pt)}$ with the vertices $-\frac{1}{2}\frac{h^{ab}}{M_{Pl}}\partial_a\psi\partial_b\psi$ and $-\frac{1}{2}\frac{l^{cd}}{M_f}\partial_c\psi\partial_d\psi$ respectively

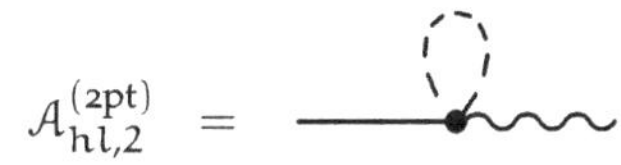

Fig. 6.19 Two point contribution $\mathcal{A}^{(2pt)}_{hl,2}$ with the vertex $\frac{1}{4}(h^{ab}[l]) + l^{ab}[h])\partial_a\psi\partial_b\psi$

$$\begin{aligned}\mathcal{A}^{(2pt)}_{hl,1} &= \frac{1}{4}(2)(3J_{M_\psi,1})\frac{1}{24}\frac{M_\psi^4}{M_{Pl}M_f}([h][l]+2[hl]) \\ &= \frac{1}{16}\frac{M_\psi^4}{M_{Pl}M_f}([h][l]+2[hl])J_{M_\psi,1}\end{aligned} \tag{6.135}$$

Now, for the last diagram we need the explicit mixing between h and l in the coupling with the matter field. If we expand the effective metric to first order in hl, then the interaction we have to consider is given by $\frac{1}{4}(h^{ab}[l]) + l^{ab}[h])\partial_a\psi\partial_b\psi$. This diagram on the other hand gives rise to a contribution of the form (Fig. 6.19)

$$\begin{aligned}\mathcal{A}^{(2pt)}_{hl,2} &= (-1)\frac{1}{4}\frac{M_\psi^4}{M_{Pl}M_f}\left(\frac{1}{4}[h][l]+\frac{1}{4}[h][l]\right)J_{M_\psi,1} \\ &= -\frac{1}{8}\frac{M_\psi^4}{M_{Pl}M_f}[h][l]\end{aligned} \tag{6.136}$$

Adding these two diagrams together (and taking the mirror imaged diagrams into account as well: factor of two) gives

$$\mathcal{A}^{(2pt)}_{hl} = 2(\mathcal{A}^{(2pt)}_{hl,1}+\mathcal{A}^{(2pt)}_{hl,2}) = \frac{M_\psi^4}{M_{Pl}M_f}\left(\frac{1}{8}([h][l]+2[hl])-\frac{1}{4}[h][l]\right) \tag{6.137}$$

As you can see from all these tadpole and two-point function contributions, the obtained counter terms do not seem to have any nice ghost-free structure. Thus the coupling through the effective metric (6.126) detunes the potential structure through one-loop matter contributions. The mass of the ghost introduced through this detuning can be made arbitrarily small destroying the classical theory.

The third scenario:—The third case which also needs to be considered is the case with explicit interactions between the different matter fields (for example $\chi_1\chi_2$) while they are coupled to the two metrics separately.

$$\sqrt{-g}\mathcal{L}_m(g,\chi_1)+\sqrt{-f}\mathcal{L}_m(f,\chi_2)+\left(\sqrt{-g}+\sqrt{-f}\right)\chi_1\chi_2\mu^2 \tag{6.138}$$

One can diagonalize this Lagrangian by performing a rotation

$$\begin{aligned}\chi_1 &= \cos(\theta)\phi_1 - \sin(\theta)\phi_2 \\ \chi_2 &= \sin(\theta)\phi_1 + \cos(\theta)\phi_2\end{aligned} \tag{6.139}$$

Fig. 6.20 One loop diagrams in which the the two metric perturbations do not mix and give rise to separate contributions

Fig. 6.21 One loop diagrams in which the the two metric perturbations mix

such that the new Lagrangian perturbed around euclidean flat metric (6.104)

$$\mathcal{L} = -\frac{1}{2}(\partial\,\phi_1)^2 - \frac{1}{2}M_{\phi_1}\,\phi_1^2 - \frac{1}{2}(\partial\,\phi_1)^2 - \frac{1}{2}M_{\phi_2}\,\phi_2^2 + \mathcal{O}(\frac{1}{M_{Pl}}, \frac{1}{M_f}) + \cdots \quad (6.140)$$

After performing this diagonalization it becomes clear that the third case corresponds to the first one with the difference that the mass spectrum M_{χ_1} and M_{χ_2} change to $M_{\phi_1}(M_{\chi_1}, M_{\chi_2}, \mu)$ and $M_{\phi_2}(M_{\chi_1}, M_{\chi_2}, \mu)$ in a non-trivial way.

This was so far concerning only the matter loops. Of course, similarly as in massive gravity, we also need to compute the one-loop quantum corrections coming from the graviton loops. In contrast to the massive gravity case, we will now also have one loop diagrams in which both metrics come in mixed. Besides the diagrams of the form where the two metrics appear separately, we will also have mixed diagrams as in the Figs. 6.20 and 6.21.

It is out of scope of this thesis to study all these contributions. We expect that the separate diagrams will give rise to detuning of the potential interactions as in massive gravity even though some cancellations might be possible. Since the $f_{\mu\nu}$ is dynamical, we have one full diffeomorphism invariance which might give rise to a better behaviour at the quantum level.

6.4 Summary and Critics

In a theory of gravity both the mass and the structure of the graviton potential are fixed by phenomenological and theoretical constraints. While in GR this tuning is protected by covariance, such a symmetry is not present in massive gravity. Nevertheless, the non-renormalization theorem present in theories of massive gravity implies that these tuning are technically natural (de Rham et al. 2012; Nicolis and Rattazzi 2004), and hence do not rely on the same fine-tuning as for instance setting the cosmological constant to zero. The emergent decoupling limit non-renormalization property, as we have seen, allows one to estimate the magnitude of quantum corrections to various parameters defining the full theory beyond any limit. In particular, one can show that setting an arbitrarily small graviton mass is technically natural. The significance of the decoupling limit theory is hard to overestimate: it unambiguously determines all the physical dynamics of the theory at distances $\Lambda_3^{-1} \lesssim r \lesssim m^{-1}$, essentially capturing all physics at astrophysical and cosmological scales. Technical naturalness, along with a yet stronger non-renormalization theorem provide perfect predictivity of the theory, enforcing quantum corrections to play essentially no role at these scales. Moreover, dictated by various physical considerations, one frequently chooses to set certain relations between the two free parameters of the theory, α_3 and α_4. Non-renormalization in this case means, that such relative tunings of parameters, along with any physical consequences that these tunings may have, are not subject to destabilization via quantum corrections in the decoupling limit. Furthermore, we have explored the stability of the graviton potential further by looking at loops of matter and graviton, assuming a covariant coupling to matter (see for instance de Rham et al. (2011a) for a discussion of the natural coupling to matter and its stability).

When integrating out externally coupled matter fields, we have shown explicitly that the only potential contribution to the one-loop effective action is a cosmological constant, and the special structure of the potential is thus unaffected by the matter fields at one-loop.

For graviton loops, on the other hand, the situation is more involved—they *do* change the structure of the potential, but in a way which only becomes relevant at the Planck scale. Nevertheless, the Vainhstein mechanism that resolves the vDVZ discontinuity relies on a classical background configuration to exceed the Planck scale (i.e. $g_{\mu\nu} - \delta_{\mu\nu} \gtrsim 1$), without going beyond the regime of validity of the theory. A naïve perturbative estimate would suggest that on top of such large background configurations, the mass of the ghost could be lowered well below the Planck scale. However, this perturbative argument does not take into account the same Vainshtein mechanism that suppresses the vDVZ discontinuity in the first place. To be consistent we have therefore considered a non-perturbative background and have shown that the one-loop effective action is itself protected by a similar Vainshtein mechanism which prevents the mass of the ghost from falling below the Planck scale, even if the background configuration is large.

The simplicity of the results presented in this study rely on the fact that the coupling to matter is taken to be covariant and that at the one-loop level virtual gravitons and matter fields cannot mix. Thus at one-loop virtual matter fields remain unaware of the graviton mass. This feature is lost at higher loops where virtual graviton and matter fields start mixing.

Higher order loops are beyond the scope of this paper, but will be investigated in depth in de Rham, Heisenberg, and Ribeiro (de Rham et al.). In this follow-up study, we will show how a naïve estimate would suggest that the two-loop graviton-matter mixing can lead to a detuning of the potential already at the scale $M_{Pl}(m/M)^2$, where m is the graviton mass and M is the matter field mass. If this were the case, a matter field with $M \sim \Lambda_3$ would already bring the mass of the ghost down to Λ_3 which would mean that the theory could never be taken seriously beyond this energy scale (or its redressed counterpart, when working on a non-trivial background). However, this estimate does not take into account the very special structure of the ghost-free graviton potential which is already manifest in its decoupling limit. Indeed, in ghost-

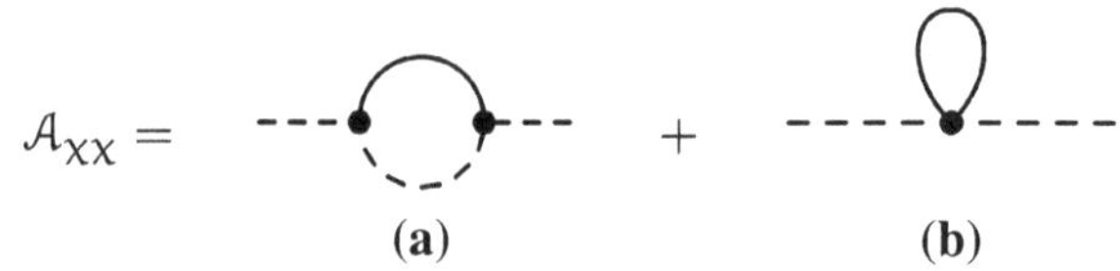

Fig. 6.22 Contribution to the scalar field two-point function from graviton/matter loops

free massive gravity the special form of the potential leads to interesting features when mixing matter and gravitons in the loops.

To give an idea of how this mixing between gravitons and matter arises, let us consider the one-loop contribution to the scalar field two-point function depicted in Fig. 6.22 if the scalar field does not have any self-interactions. We take the external leg of the scalar field to be on-shell, i.e. with momentum q_a satisfying $\delta^{ab}q_a q_b + M^2 = 0$.[9]

For the purpose of this discussion, it is more convenient to work in terms of the field χ directly rather than the redefined field ψ. First, diagram (a) gives rise to a contribution proportional to

$$\mathcal{A}^{(a)}_{\chi\chi} \propto \frac{1}{M_{Pl}^2}\int d^4k \frac{f^{(m)}_{abcd}(k)\, q^a p^b q^c p^d}{(k^2+m^2)(p^2+M^2)}, \tag{6.141}$$

where k is the momentum of the virtual graviton running in the loop of diagram (a). By momentum conservation, the momentum p of the virtual field χ in the loop is then $p_a = q_a - k_a$. Applying the on-shell condition for the external legs, $q^2 + M^2 = 0$, we find

$$\mathcal{A}^{(a)}_{\chi\chi} \propto \frac{1}{m^4 M_{Pl}^2}\int d^4k \left((k.q)^2 + m^2 q^2\right) \equiv 0. \tag{6.142}$$

in dimensional regularization. So this potentially 'problematic' diagram that mixes virtual matter and gravitons (which could a priori scale as m^{-4}) leads to no running when the *external* scalar field is on-shell. In other words, at most this diagram can only lead to a running of the wave-function normalization and is thus harmless (in particular it does not affect the scalar field mass, nor does it change the 'covariant' structure of the scalar field Lagrangian).

Second, we can also consider the contribution from a pure graviton-loop in diagram (b). Since only the graviton runs in that loop, the running of the scalar field mass arising from that diagram is at most $\delta M^2 = \frac{m^2}{M_{Pl}^2}M^2 \ll M^2$ and is therefore also harmless. As a consequence we already see in this one-loop example how the mixing between the virtual graviton and scalar field in the loops keeps the structure of the matter action perfectly under control.

[9] Technically, in Euclidean space this means that the momentum is complex, or one could go back to the Lorentzian space-time for the purposes of this calculation, but these issues are irrelevant for the current discussion. Moreover, note that the on-shell condition is only being imposed for the external legs, and *not* for internal lines.

Even if we can not provide the exact behavior of the two loop contributions in which now both the graviton as well as the matter fields propagate in this thesis, we comment on the contributions with the highest and the sub-highest scaling in the inverse of the graviton mass which cancel exactly due to the antisymmetric structure of the interactions. For this purpose lets have a closer look to the contribution of the tadpole with two internal graviton propagators (Fig. 6.23). For instance lets consider for the first vertex the interaction coming from the potential term which has the antisymmetric structure $V = \mathcal{E}^{abcd}\mathcal{E}^{a'b'c'}{}_{d}\, v_{aa'}v_{bb'}v_{cc'}$. We can contract one of the spin-2 field with an external $v_{\mu\nu}$ leg coming out of this vertex while letting the other two spin-2-fields from this vertex run in the loop with momenta k and contract them at the second and third vertex with the two spin-2 fields coming from the couplings $v_{\alpha\alpha'}T^{\alpha\alpha'}$ and $v_{\beta\beta'}T^{\beta\beta'}$. Now, if each internal propagator comes at least with k^4/M^4 than the contribution of this vertex to the graph with the most negative powers of the graviton mass would be m^{-8}

$$\mathcal{A} \propto \int \frac{d^4k}{(2\pi)^4}\frac{e^{ik_u(x_1^u - x_2^u)}}{k^2+m^2} www_{\mu\nu}\mathcal{E}^{abcd}\mathcal{E}^{a'b'c'}{}_{d} \times \frac{k_a k_{a'}}{m^2}\frac{k_b k_{b'}}{m^2}\frac{k_\alpha k_{\alpha'}}{m^2}\frac{k_\beta k_{\beta'}}{m^2} T^{\alpha\alpha'}T^{\beta\beta'}\dots \tag{6.143}$$

Now, it is a trivial observation that this contribution cancels due to the antisymmetric structure of the vertex. The same happens to the contribution with the m^{-6} scaling. For the 2-point function consider the following diagram with three internal graviton propagators (Fig. 6.24). These diagrams could in principle yield contributions with the bad scalings m^{-12}, m^{-10}, m^{-8} and m^{-6}. However, we have checked explicitly that these contributions are exactly zero as well. Again, the nice antisymmetric structure of the potential prohibits the appearance of these scalings.

$$\mathcal{E}^{abcd}\mathcal{E}^{a'b'c'}{}_{d}\frac{k_a k_{a'} k_b k_{b'} k_c k_{c'}}{m^6}\frac{k^6}{m^6} = 0 \tag{6.144}$$

We observe the same cancelation for the most negative power of the graviton mass as well when we include the interactions coming from the Einstein Hilbert action,

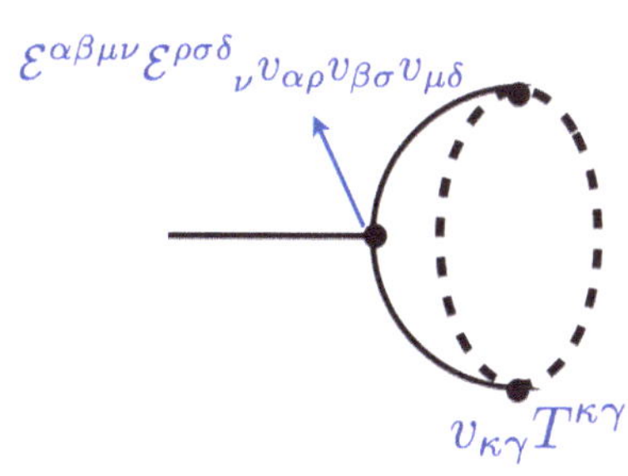

Fig. 6.23 Tadpole function with 2-loops

Fig. 6.24 2-point function with 2-loops

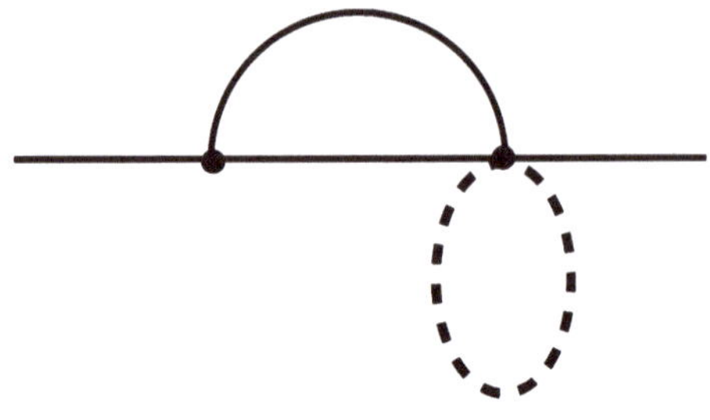

not just the potential term. At first sight this might look surprising. But if we recall that the Einstein–Hilbert term can also be written as

$$\sqrt{-g}R = 4\sqrt{-g}\mathcal{E}^{abcd}\mathcal{E}^{a'b'c'd'}g_{aa}g_{bb'}R_{cdc'd'} \tag{6.145}$$

Exactly in the same way as it was the case for the potential term, the antisymmetric Levi-Civita tensor prohibits the appearance of the highest and sub highest negative power of the graviton mass. Contributions of the form

$$\mathcal{A} \propto \int \frac{d^4k}{(2\pi)^4}\frac{e^{ik_u(x_1^u - x_2^u)}}{k^2 - i\varepsilon} v_{cd}\mathcal{E}^{abcd}\mathcal{E}^{a'b'c'd'} \\ \frac{k_a k_{a'}}{m^2}\frac{k_b k_{b'}}{m^2}\frac{k_\alpha k_{\alpha'}}{m^2}\frac{k_\beta k_{\beta'}}{m^2} k_{c'}k_{d'}T^{\alpha\alpha'}T^{\beta\beta'} \tag{6.146}$$

cancel. Since the Riemann tensor contains two derivatives applied on the vielbein, the contributions from the Einstein Hilbert action encloses two more k's besides the ones coming from the propagator and hence reinforces the argumentation based on the antisymmetry of the Levi-Civita tensors (of course this is not the case if the Riemann tensor contributes to $\mathcal{O}(1)$ order only). For a diagram with two internal graviton propagators the worst scalings are $\frac{k^8}{m^8}k^2$ and $\frac{k^6}{m^6}k^2$ and the index structure of the k's (with now two more k's) makes that these contributions vanish.

Fig. 6.25 n-point function with 2 Loops

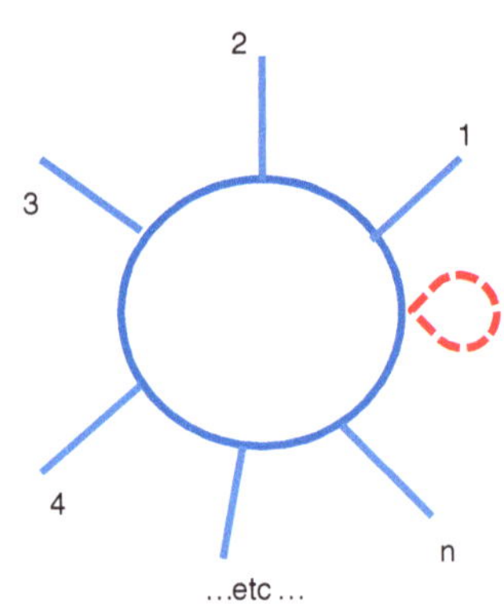

We can generalize the above observation to any n-point function (Fig. 6.25). Again the n-point function with the highest number of internal graviton propagators n+1 will contribute with the most negative power of $m^{-4(n+1)}$, which again cancels exactly. The reason for that is simple. We already know that the cubic vertex from the potential term and also from the Einstein–Hilbert term gives zero lo leading order and m^2 scaling to second leading order. Therefore for the n-point function, each vertex cannot contribute with more than m^{-2}, meaning that the divergence is at worst like $m^{-2(n+1)}$ rather than $m^{-4(n+1)}$. Even if this argumentation is based on cubic vertices, the same reasoning applies to quartic and higher dimensional vertices in v. Each internal propagator comes at least with k^2/m^2. In the case in which each propagator contributes with k^2/m^2 would give rise to $(k/m)^{2n}$ contributions in the vertex which is fully symmetric and cancel due to the antisymmetric structure of the potential term with the Levi-Civita tensors.

We are able to explain why the quantum corrections with the worst scaling of the inverse graviton mass cancel, nevertheless at 2-loops there could be still some contributions in which the detuning of the potential interactions occur at a lower scale then at the one-loop level even if the inverse scaling does not happen. Therefore, it is an indispensable task to push our analysis beyond the one-loop contributions. Even if there is no apparent symmetry to protect the specific structure of the massive gravity potential, computing explicitly the quantum corrections at one and two-loops might provide us with an iterative scheme to show the absence of ghosts below the Planck scale at all order of loops.

References

Berezhiani L, Chkareuli G, de Rham C, Gabadadze G, Tolley AJ (2012) On black holes in massive gravity. Phys Rev D 85:044024

Buchbinder IL, de Berredo-Peixoto G, Shapiro IL (2007) Quantum effects in softly broken gauge theories in curved space-times. Phys Lett B 649:454–462. doi:10.1016/j.physletb.2007.04.039

Buchbinder IL, Pereira DD, Shapiro IL (2012) One-loop divergences in massive gravity theory. Phys Lett B 712:104–108. doi:10.1016/j.physletb.2012.04.045

Chamseddine AH, Mukhanov V (2011) Massive gravity simplified: a quadratic action. JHEP 1108:091

de Rham C, Gabadadze G (2010) Generalization of the Fierz-Pauli action. Phys Rev D 82:044020. doi:10.1103/PhysRevD.82.044020

de Rham C, Gabadadze G, Heisenberg L, Pirtskhalava D (2012) Non-renormalization and naturalness in a class of scalar-tensor theories. Phys Rev D 87:085017

de Rham C, Gabadadze G, Tolley AJ (2011a) Helicity decomposition of ghost-free massive gravity. JHEP 1111:093. doi:10.1007/JHEP11(2011)093

de Rham C, Gabadadze G, Tolley AJ (2011b) Resummation of massive gravity. Phys Rev Lett 106:231101. doi:10.1103/PhysRevLett.106.231101

de Rham C, Heisenberg L, Ribeiro RH Quantum corrections in massive gravity at higher loop: mixing matter with gravitons (to appear)

Deffayet C, Jacobson T (2012) On horizon structure of bimetric spacetimes. Class Quant Grav 29:065009. doi:10.1088/0264-9381/29/6/065009

Deffayet C, Mourad J, Zahariade G (2013) A note on 'symmetric' vielbeins in bimetric, massive, perturbative and non perturbative gravities. JHEP 1303:086
Fasiello M, Tolley AJ (2013) Cosmological stability bound in massive gravity and bigravity. JCAP 1301:032
Fierz M, Pauli W (1939) On relativistic wave equations for particles of arbitrary spin in an electromagnetic field. Proc Roy Soc Lond A 173:211–232
Gabadadze G, Hinterbichler K, Pirtskhalava D, Shang Y (2013) On the potential for general relativity and its geometry. Phys Rev D 88:084003
Groot SN, Marco P, Matthew S (2007) Nonlinear properties of vielbein massive gravity. Eur Phys J C 51:741–752
Hassan SF, Rosen RA (2011) On non-linear actions for massive gravity. JHEP 1107:009. doi:10.1007/JHEP07(2011)009
Hassan SF, Rosen RA (2012a) Bimetric gravity from ghost-free massive gravity. JHEP 1202:126. doi:10.1007/JHEP02(2012)126
Hassan SF, Rosen RA (2012b) Resolving the ghost problem in non-linear massive gravity. Phys Rev Lett 108:041101.10.1103/PhysRevLett.108.041101
Hinterbichler K, Rosen RA (2012) Interacting Spin-2 fields. JHEP 1207:047. doi:10.1007/JHEP07(2012)047
Koyama K, Niz G, Tasinato G (2011a) Analytic solutions in non-linear massive gravity. Phys Rev Lett 107:131101. doi:10.1103/PhysRevLett.107.131101
Koyama K, Niz G, Tasinato G (2011b) Strong interactions and exact solutions in non-linear massive gravity. Phys Rev D 84:064033. doi:10.1103/PhysRevD.84.064033
Nicolis A, Rattazzi R (2004) Classical and quantum consistency of the DGP model. JHEP 0406:059. doi:10.1088/1126-6708/2004/06/059
Ondo NA, Tolley AJ (2013) Complete decoupling limit of ghost-free massive gravity. JHEP 1311:059
Park M (2011) Quantum aspects of massive gravity. Class Quant Grav 28:105012
't Hooft G, Veltman MJG (1974) One loop divergencies in the theory of gravitation. Ann Poincare Phys Theory A20:69–94
Vainshtein AI (1972) To the problem of nonvanishing gravitation mass. Phys Lett B 39:393–394. doi:10.1016/0370-2693(72)90147-5
van Dam H, Veltman MJG (1970) Massive and massless Yang-Mills and gravitational fields. Nucl Phys B 22:397–411
Zakharov VI (1970) Linearized gravitation theory and the graviton mass. JETP Lett 12:312

Part IV
Summary

Chapter 7
Summary and Outlook

7.1 Summary

The main focus of this thesis was the study of theoretical as well as phenomenological consequences derived from the existence of a graviton mass within the ghost-free theory of Massive Gravity, which constitutes a 2-parameter family of potentials, known under the name of dRGT theory. First of all, we studied the impact of Massive Gravity on cosmological contexts and obtained constraints on the parameters of the theory based on purely theoretical grounds. To be more precise we studied the self-accelerating and degravitating solutions in the decoupling limit of Massive Gravity and put constraints on the two free parameters of the theory from instability conditions (like the absence of ghost and Laplacian instabilities) in the cosmological evolution. The self-accelerating solution we found was indistinguishable from a cosmological constant, since the fluctuations of the extra scalar degree of freedom of the massive graviton sourcing self-acceleration do not couple to the fluctuations of the helicity-2 field and so to the matter fields. We were also able to show that Massive Gravity can screen an arbitrarily large cosmological constant in the decoupling limit without giving rise to any ghost instability. Unfortunately, the allowed value of the vacuum energy that can be screened without being in conflict with observations is fairly low. Even if the screening of a large cosmological constant pushes the Vainshtein radius to smaller scales, to our knowledge it is the first time that an explicit model can present a way out from Weinberg's no-go theroem.

Furthermore, starting from the decoupling limit we constructed a proxy theory to Massive Gravity which gave rise to non-minimally coupled scalar-tensor interactions as an example of a subclass of Horndeski theories. Even if the Horndeski scalar-tensor interactions have been studied in the literature before, the novelty in our study was that these very specific subclass interactions descend from the very specific structure of the decoupling limit of Massive Gravity. We studied the self-accelerating and degravitating solutions in this proxy theory as well. We found stable self-accelerating solutions and followed the cosmological evolution of the extra scalar degree of freedom. In difference to the self-accelerating solution found in the decoupling limit

L. Heisenberg, *Theoretical and Observational Consistency of Massive Gravity*,
Springer Theses, DOI 10.1007/978-3-319-18935-2_7

of Massive Gravity, the one found in the proxy theory through covariantization differs from a pure cosmological constant since the perturbations of the extra degree of freedom do couple to the matter fields. Motivated by this result, we studied the change in the linear structure formation. Unfortunately, we were not able to find any degravitating solutions with an analogue ansatz as in the decoupling limit.

Additionally, Galileon models represent a class of effective field theories that naturally arise in the decoupling limit of theories of massive gravity. Even though Galileon models offer interesting phenomenology, they suffer from superluminal propagation. We have shown that superluminal propagating solutions for multi-galileon theories is an unavoidable feature, unlike previously claimed in the literature.

After studying the constraints on the parameters of the theory based on stability condition, the natural question arises as to whether the tuning of the introduced parameters themselves are subject to strong renormalization by quantum loops. Therefore we pushed the stability analysis from pure classical to the quantum level by addressing the fundamental question of technical naturalness of the parameters at the quantum level. We started the analysis with the decoupling limit and showed how the non-renormalization theorem implies that the graviton mass and the free parameters of the theory receive no quantum corrections. This non-renormalization works in an exact analog way as the non-renormalization for the Galileon fields. The interactions are safe from quantum corrections because of the antisymmetric structure of the interactions. This is a very nice exact result which hold for any number of quantum loops. Beyond the decoupling limit the quantum corrections are present but suppressed in such a way that the graviton mass can only be renormalized by an amount proportional to itself, proving the technical naturalness of the small graviton mass. This provides an explicit realization of the 't Hooft naturalness argument to the case of Massive Gravity. We then explicitly studied the stability of the graviton potential under the quantum corrections coming from loops of matter and gravitons in order to confirm 't Hooft naturalness argument. Hereby we assumed a covariant coupling to matter. We found that one-loop contributions from the externally coupled matter fields yield a contribution to the one-loop effective action in form of a cosmological constant and that therefore the special structure of the potential is unaffected by the matter fields. On the contrary, the one-loop contributions from the gravitons themselves destabilize the special structure of the potential. Nevertheless, this detuning is irrelevant below the Planck scale. Furthermore, we showed that even in the case of large background configuration, the Vainshtein mechanism redresses the one-loop effective action so that the detuning remains irrelevant below the Planck scale. Our work was one of the first attempts at invoking the Vainshtein mechanism at the quantum level.

7.2 Outlook

There are still very interesting and promising roads to take within the field of Massive Gravity and multi-metric gravities which can be divided into observational and more theoretical considerations rending the field interesting for different communities.

7.2.1 Theoretical Concepts

An interesting follow-up investigation of the quantum corrections in Massive Gravity is the question whether or not our results for one-loop corrections remain valid at higher order loops. At the one loop level there was no mixing between graviton and scalar propagators within loops since either gravitons or matter fields could run in the loops. As a consequence the virtual matter fields remained unaware of the graviton mass. But starting from two loops this mixing will play an essential role and it becomes a compulsory task to study whether or not this higher order loop graviton-matter mixing will detune the potential at a different scale and correspondingly yield a smaller mass for the ghost, which would render the theory sick. Could it be that the special form of the potential forces the mass of the ghost never to be less than the Planck mass? Could loops with graviton-matter mixing give rise to an inverse power scaling of the graviton mass? Could it be that the same very specific form of the graviton potential arranges for the infinite number of loop diagrams to resume into interactions which stay in the weakly-coupled regime? Can the theory be UV-completed a la Wilsonian or forced to have an alternative approach as classicalization? Additionally, the propagation of superluminal fluctuations in these theories signals the failure of having a Wilsonian UV-completion. Curiously, the Vainshtein mechanism and the superluminal propagation seem to appear hand in hand if one imposes trivial asymptotic conditions at infinity. Thus, it is crucial to investigate the deep physical relations between all these phenomena which are up to date still mysterious. All these questions are worth the effort.

Coming back to the original motivation of studying Massive Gravity, the CC problem remains still undressed in the full theory. Even if the decoupling limit of Massive Gravity fails to degravitate an arbitrary large CC in order to make the Vainshtein mechanism work, an appealing follow-up project which is worth considering, is the investigation of degravitating solutions in the full theory. Surprisingly degravitating solutions of Massive Gravity have been left aside so far in the literature. Could it be that in the full theory there exists a cosmological scenario in which the degravitation of the vacuum energy is not in conflict with the Vainshtein mechanism or that the degravitating even takes place before the Vainshtein mechanism starts to be at work? Another possible project is to investigate degravitating solutions with varying mass or in bimetric gravity theories. Bimetric gravity theories provide a rich phenomenology in the cosmological context. Stable self-accelerating solutions have been found in the literature. Therefore it is a promising road to study in more detail these potential effects that could give rise to cosmological signatures and that could help discriminating the theory as well as constraining its parameters.

It is very interesting that Galileon type of theories and Massive Gravity can be regarded as effective descriptions of braneworld constructions or compactifications in higher-dimensional frameworks. The fact that some of the standard model of elementary particles problems, e.g. the Higgs hierarchy problem, can also be attacked by resorting to extra dimensions makes it even more intriguing. Thus, the two most important hierarchy problems, namely the cosmological constant and the Higgs mass problems, find potential solutions in scenarios with extra-dimensions. This is very suggestive and therefore finding ways to tackle both puzzles in a unified manner from higher-dimensional setups deserves further investigations.

7.2.2 Observational Concepts

The confrontation of Massive Gravity and bimetric gravity theories with cosmological observations is a crucial ingredient in testing these theories. A natural starting point is the study of the background evolution and to constrain the parameters using the distance redshift relation from Supernovae, the distance priors method from CMB and BAO measurements. The cosmological observations with the most constraining power will most likely be the spectrum of temperature fluctuations of the CMB which requires in addition a dark matter spectrum and a growth function. Thus, for this purpose one of the publicly available numerical codes, like CAMB or CLASS, needs to be modified to solve the whole system of Boltzmann equations together with Einstein equations so that we can confront the model to CMB observations and obtain the corresponding confidence regions for the theory parameters using Monte-Carlo Markov-chains.

At the linear level, the CMB temperature anisotropies provide an excellent tool to probe the evolution of the inhomogeneous perturbations in the universe. However, these are not sufficiently powerful to disentangle certain degeneracies that exist between different cosmological models or modified gravities. The non-linear evolution is in fact responsible for the formation of the structures that we observe in the universe and, in order to fully understand how the structures in our universe are formed, we need to resort to methods beyond the linear regime. This is so because the gravitational processes involved in the formation of galaxies, clusters, etc. from the primordial seeds generated during inflation are highly non-linear. The existence of a non-vanishing mass for the graviton naturally leads to a modification of Newton's potential, changing in this way the gravitational interaction between particles. Thus, the clustering that takes place in a universe with a massive graviton will differ from the standard one. Moreover, since not only the gravitational forces, but also the cosmological expansion affects the structure formation, another effect to be considered is how the different Hubble expansion rate produced by a massive graviton could impact on it. In order to perform this analysis, we need to start by studying cosmological perturbations within the context of Massive Gravity. In particular, we would study the spherical collapse to further proceed to the question of how the density contrast of matter fluctuations is modified with respect to the standard model. Once this preliminary study has been accomplished, the implementation of the corresponding modifications in an N-body code to run simulations of structure formation will allow a very important step in testing the theory because of the non-linearity of the gravitational equations. There is a series of features that cannot be properly accounted for by resorting to cosmological perturbation theory alone, but the solution of the entire system of equations is needed. Since the existence of a mass for the graviton makes the gravitational interaction weaker on intermediate and large scales with respect to General Relativity, and the Vainshtein mechanism will play a crucial role in the gravitational collapse, we expect a different structure formation in Massive Gravity that will allow to distinguish it from other models. In any case, and given the complexity of the problem, N-body simulations should be run to see the actual features produced within the framework of Massive Gravity. The use of N-body simulations is computationally very expensive so that it is crucial to develop reliable and optimal semi-analytical codes with less computational time. In this context, it would be useful to extend the existing PINOCCHIO code to cope with modified gravity models such as Massive Gravity which differ from conventional models such as ΛCDM by a different onset of structure formation, and investigate the parameter space of Massive Gravity. We presume that Massive Gravity will significantly influence the time sequence of gravitational clustering, especially the evolution of peculiar velocities and the number density of collapsed objects in the PINOCCHIO code. Developing a code including this novel scalar non-linear interactions with the Vainshtein mechanism implemented will be a key contribution to the research field. Moreover, the development of such a code will require a deep understanding of the problem of structure formation so that we will gain, not only intuition and experience within the context of strongly coupled scalar field modifications of gravity, but also on the process of structure formation in more general setups.

EUCLID provides indispensable possibilities to study the effects of Massive Gravity in weak lensing measurements. Using Fisher-formalism for forecasting statistical errors on parameters, EUCLID weak lensing measurements will restrict the parameter space of Massive Gravity remarkably. Particularly the mass of the graviton can be determined by weak lensing. For this purpose the power spectrum as well as the growth rate are needed. Bayesian model selection will then enable us to differentiate between Massive Gravity and competing dark energy and modified gravity models. Interestingly, EUCLID will be testing a region of Massive Gravity just at the Vainshtein radius itself which is so far not very well probed.

Appendix

A.1 Specific Example Provided in Padilla et al. (2011)

Following the example provided in Padilla et al. (2011), we now take the following numerical values for the parameters $a_{m,n}$ and $b_{m,n}$: $a_{10} = 3\mu^2$, $a_{01} = -1\mu^2$, $b_{01} = 1/2\mu^2$, $a_{30} = 2\mu^2/H_0^4$, $a_{21} = -13\mu^2/H_0^4$, $a_{12} = 24\mu^2/H_0^4$, $a_{03} = -9\mu^2/H_0^4$ and $b_{03} = -18\mu^2/H_0^4$. Plugging these values into the expression for the sound speed yields:

$$cs_+^2(r) \sim 1 + 133.9\frac{m^2}{r^6}\frac{1}{H_0^4\mu^4} \tag{A.1}$$

$$cs_-^2(r) \sim 1 - 14933.9\frac{m^2}{r^6}\frac{1}{H_0^4\mu^4} \tag{A.2}$$

We thus have one mode that propagates sub-luminaly at infinity and the other mode propagating super-luminaly. It therefore seems that the example provided in Padilla et al. (2011) does not achieve the goal of avoiding superluminal propagation. At this level, this might be just due to a simple typo in the specific example provided in Padilla et al. (2011), but as we show in this thesis, there are no possible sets of quadratic and quartic bi-Galileon interactions that can prevent the propagation of one superluminal mode, even though the parameters at infinity are chosen such that bode modes propagate (sub-)luminally at far infinity.

A.2 Detailed Analysis of the Special Case in the Quartic Galileon: Dominant First Order Corrections

In this section of the appendix we will show that $\tau - \delta > 0$ when $\lambda_1^2 > \lambda_2^2$.

L. Heisenberg, *Theoretical and Observational Consistency of Massive Gravity*,
Springer Theses, DOI 10.1007/978-3-319-18935-2

We can write $\tau - \delta$ as

$$\tau - \delta = \frac{6b_{01}m}{\lambda_1^4 y_1}(1-\alpha) \times \Big[\frac{1}{8(3\alpha-1)^2}\left(a_{10}b_{01} - 2\frac{3\alpha-1}{5\alpha-1}a_{01}^2\right) + \frac{2}{(1-\alpha)(5\alpha-1)}\zeta^2 - \frac{1}{(5\alpha-1)(3\alpha-1)}\zeta a_{01}\Big]. \quad \text{(A.3)}$$

We also use the notation $\lambda_2^2 = \alpha\lambda_1^2$. Note that since $y_1, b_{01} > 0$ we must have $3\lambda_2^2 > \lambda_1^2$. Thus we have $1/3 < \alpha < 1$, the upper bound comes from our assumption that $\lambda_1^2 > \lambda_2^2$.

Now the first term in the brackets of the expression $\tau - \delta$ has the same sign as

$$a_{01}b_{01} - \epsilon a_{01}^2\,, \quad \text{(A.4)}$$

with $0 < \epsilon < 1$. This is positive because in order to avoid ghost instabilities at large distances from the source $a_{10}b_{01} - a_{01}^2 > 0$ and $a_{10}, b_{01} > 0$, (see Eq. (4.47)). Meanwhile, the second term in the brackets is manifestly positive. Finally, the third term has the sign of $-\zeta a_{01}$.

So at this point our only hope of avoiding superluminalities is to consider a choice of parameters where $-\zeta a_{01} < 0$. Now we will proceed to show that $\tau - \delta > 0$ in this case as well.

Note that in the limit $\alpha \to 1$ with everything else fixed we have

$$\tau - \delta \longrightarrow \frac{3b_{01}m\zeta^2}{\lambda_1^4 y_1} > 0\,. \quad \text{(A.5)}$$

Also in the limit $\alpha \to 1/3$ with everything else fixed we have

$$\tau - \delta \longrightarrow \frac{a_{10}b_{01}^2 m}{2\lambda_1^4 y_1}\frac{1}{(1-3\alpha)^2} > 0\,. \quad \text{(A.6)}$$

Now consider the function

$$\sigma(\alpha) = 8(1-\alpha)(5\alpha-1)(3\alpha-1)^2\Big[\frac{1}{8(3\alpha-1)^2}\left(a_{10}b_{01} - 2\frac{3\alpha-1}{5\alpha-1}a_{01}^2\right) + \frac{2}{(1-\alpha)(5\alpha-1)}\zeta^2 - \frac{1}{(5\alpha-1)(3\alpha-1)}\zeta a_{01}\Big]. \quad \text{(A.7)}$$

We can write this function in the shortened notation as $\sigma(\alpha) = \sigma_0 + \sigma_1 \alpha + \sigma_2 \alpha^2$ with

$$\sigma_0 = 2a_{01}^2 - a_{10}b_{01} + 8a_{01}\zeta + 16\zeta^2 \tag{A.8}$$

$$\sigma_1 = -8a_{01}^2 + 6a_{10}b_{01} - 32a_{01}\zeta - 96\zeta^2 \tag{A.9}$$

$$\sigma_2 = 6a_{01}^2 - 5a_{10}b_{01} + 24a_{01}\zeta + 144\zeta^2 \,. \tag{A.10}$$

The sign of $\sigma(\alpha)$ is the same as the sign of $\tau - \delta$ in the regime $1/3 < \alpha < 1$. Note that $\sigma(1/3), \sigma(1) > 0$ using the limits above.

Note that $\sigma_0, \sigma_1, \sigma_2$ do not have a definite sign, because $a_{01}\zeta, \zeta^2 > 0$, but $a_{01}^2 - a_{10}b_{01} < 0$. Therefore we need to investigate the behaviour of this function $\sigma(\alpha)$ in more detail.

Being a quadratic function $\sigma(\alpha)$ has a single critical point (either corresponding to a maximum or a minimum) α_{crit}. Given that $\sigma(1/3), \sigma(1) > 0$, we can avoid superluminalities if and only if $1/3 < \alpha_{\text{crit}} < 1$ and simultaneously $\sigma(\alpha_{\text{crit}}) < 0$.

Computing $\frac{d\sigma(\alpha)}{d\alpha} = 0$ yields for the critical point α_{crit}

$$\alpha_{\text{crit}} = -\frac{\sigma_1}{2\sigma_2} = \frac{4a_{01}^2 - 3a_{10}b_{01} + 16a_{01}\zeta + 48\zeta^2}{\sigma_2} \,. \tag{A.11}$$

Plugging this back into the expression for $\sigma(\alpha_{\text{crit}})$ gives the following expression

$$\sigma(\alpha_{\text{crit}}) = 4(a_{10}b_{01} - a_{01}^2)\frac{a_{01}^2 + 8a_{01}\zeta + 16\zeta^2 - a_{10}b_{01}}{\sigma_2} \,. \tag{A.12}$$

It is useful to consider

$$1 - \alpha_{\text{crit}} = 2\frac{a_{01}^2 + 4a_{01}\zeta + 48\zeta^2 - a_{10}b_{01}}{\sigma_2} \,. \tag{A.13}$$

If $\alpha_{\text{crit}} < 1$ then this is positive. Similarly

$$\alpha_{\text{crit}} - \frac{1}{3} = \frac{1}{3}\frac{6a_{01}^2 + 24a_{01}\zeta - 4a_{10}b_{01}}{\sigma_2} \,. \tag{A.14}$$

If $\alpha_{\text{crit}} > 1/3$ then this is positive.[1]

We will now show that we cannot simultaneously satisfy all the criteria that we need to satisfy to avoid superluminalities. We consider four cases which will exhaust all possibilities:

[1] When we write num of $\sigma(\alpha_{\text{crit}})$, we mean $a_{01}^2 + 8a_{01}\zeta + 16\zeta^2 - a_{10}b_{01}$ by that, i.e. we are ignoring the uninteresting factor of $4(a_{10}b_{01} - a_{01}^2) > 0$.

Case 1: $\sigma_2 = 0$

In this case we have

$$\sigma(\alpha) = \underset{0}{\sigma} + \underset{1}{\sigma}\alpha$$

Since $\sigma(1/3), \sigma(1) > 0$ we know that $\sigma(\alpha) > 0$ in the whole interval $1/3 < \alpha < 1$.

Case 2: $\sigma_2 < 0$

Consider $\sigma(\alpha_{\text{crit}})$. If we assume that σ_2 is negative, then we can avoid superluminalities if and only if the numerator of $\sigma(\alpha_{\text{crit}})$ is positive.

However the condition that σ_2 is negative means that $a_{01}b_{01} > \frac{6}{5}a_{01}^2 + \frac{24}{5}a_{01}\zeta + \frac{144}{5}\zeta^2$, which implies that

$$\text{num of } \sigma(\alpha_{\text{crit}}) = a_{01}^2 + 8a_{01}\zeta + 16\zeta^2 - a_{10}b_{01} < -\frac{1}{5}(a_{01} - 8\zeta)^2 . \qquad \text{(A.15)}$$

So we cannot avoid superluminalities in this case either.

Case 3: $\sigma_2 > 0, \zeta > 0$

Again we consider $\sigma(\alpha_{\text{crit}})$. We now assume that σ_2 is positive, so we need to check if numerator of $\sigma(\alpha_{\text{crit}})$ can be negative if we also assume that $\alpha_{\text{crit}} < 1$, i.e. $a_{10}b_{01} < a_{01}^2 + 4a_{01}\zeta + 48\zeta^2$, and also that $\alpha_{\text{crit}} > 1/3$, i.e. $a_{01}b_{01} < \frac{3}{2}a_{01}^2 + 6a_{01}\zeta$.

The inequality $\alpha_{\text{crit}} < 1$ tells us that

$$\text{num of } \sigma(\alpha_{\text{crit}}) = a_{01}^2 + 8a_{01}\zeta + 16\zeta^2 - a_{10}b_{01} > 4a_{01}\zeta - 32\zeta^2 = 4\zeta(a_{01} - 8\zeta) \qquad \text{(A.16)}$$

and the inequality $\alpha_{\text{crit}} > 1/3$ tells us that

$$\text{num of } \sigma(\alpha_{\text{crit}}) = a_{01}^2 + 8a_{01}\zeta + 16\zeta^2 - a_{10}b_{01} > -\frac{1}{2}a_{01}^2 + 2a_{01}\zeta + 16\zeta^2 \qquad \text{(A.17)}$$

Now let's take $\zeta > 0$. The first inequality then implies we need $a_{01} - 8\zeta < 0$ to avoid superluminalities. So we set $a_{01} = 8\zeta\epsilon$ for $0 < \epsilon < 1$ (if $\epsilon < 0$ then $-a_{01}\zeta > 0$). Then the second inequality becomes

$$\text{num} > 16\zeta^2(1 + \epsilon - 2\epsilon^2) = 16\zeta^2(1 - \epsilon)(1 + 2\epsilon) > 0 \qquad \text{(A.18)}$$

So also in this case we are forced to have superluminalities.

Case 4: $\sigma_2 > 0, \zeta < 0$

Now we take $\zeta < 0$. The first inequality then implies we need $a_{01} - 8\zeta > 0$ to avoid superluminalties. However note that both a_{01} and ζ are negative.

So we set $a_{01} = 8\zeta\epsilon$ for $0 < \epsilon < 1$. Then the argument is exactly the same as above, and that concludes our set of possibilities. In conclusion there is no possible way to avoid superluminalities near the origin, even if one had been so lucky as to live in a theory with specifically tuned coefficients for which the first order corrections near the origin vanished. Our result is thus generic: superluminalities are always present near the origin if the field is to be trivial at infinity and stable both at small and large distances.

A.3 Dimensional Regularization

For the one-loop diagrams we required the dimensional regularization technique to obtain the quantum corrections. A recurrent integral which appears in our calculations is of the form

$$J_{\tilde{m},n} = \frac{1}{\tilde{m}^4} \int \frac{d^4k}{(2\pi)^4} \frac{k^{2n}}{\left(k^2+\tilde{m}^2\right)^n}, \tag{A.19}$$

where $\tilde{m}$ is a placeholder for whichever mass appears in the propagator. By symmetry we have

$$\frac{1}{\tilde{m}^4} \int \frac{d^4k}{(2\pi)^4} \frac{k^{2(n-j)}k_{\alpha_1}\cdots k_{\alpha_{2j}}}{\left(k^2+\tilde{m}^2\right)^n} = \frac{1}{2^j(j+1)!}\delta_{\alpha_1\cdots\alpha_{2j}} J_{\tilde{m},n}, \tag{A.20}$$

with the generalized Kronecker symbol,

$$\delta_{\alpha_1\cdots\alpha_{2j}} = \delta_{\alpha_1\alpha_2}\delta_{\alpha_3\cdots\alpha_{2j}} + \left(\{\alpha_2\} \leftrightarrow \{\alpha_3,\ldots,\alpha_{2j}\}\right). \tag{A.21}$$

We also note that

$$J_{\tilde{m},n} = \frac{n(n+1)}{2} J_{\tilde{m},1}. \tag{A.22}$$

We do not need to express $J_{\tilde{m},1}$ explicitly in dimensional regularization, but can simply rely on these different relations to show how different diagrams repackage into a convenient form. It suffices to know that $J_{\tilde{m},1}$ contains the logarithmic divergence in $\tilde{m}$, which represents the running in renormalization techniques.

Reference

Padilla A, Saffin PM, Zhou S-Y (2011) Bi-galileon theory II: phenomenology. JHEP 1101:099. doi:10.1007/JHEP01(2011)099

MIX
Papier aus verantwortungsvollen Quellen
Paper from responsible sources
FSC® C105338

If you have any concerns about our products,
you can contact us on
ProductSafety@springernature.com

In case Publisher is established outside the EU,
the EU authorized representative is:
Springer Nature Customer Service Center GmbH
Europaplatz 3, 69115 Heidelberg, Germany

Printed by Libri Plureos GmbH
in Hamburg, Germany